Patrick Glöckner
Tunja Jung
Susanne Struck
Katia Studer

Radiation Curing

Coatings and Printing Inks

Technical Basics, Applications and Trouble Shooting

Glöckner/Jung/Struck/Studer: Radiation Curing for Coatings and Printing Inks

ISBN: 978-3-86630-907-4

Cover picture: Urs Läuppi

Patrick Glöckner, Tunja Jung, Susanne Struck, Katia Studer
Radiation Curing for Coatings and Printing Inks: Technical Basics and Applications
Hannover: Vincentz Network, 2008
(European Coatings Tech Files)
ISBN 3-86630-907-4
ISBN 978-3-86630-907-4

Please ask for our book catalogue
Vincentz Network, Plathnerstr. 4c, 30175 Hannover, Germany
T (202) 684-6630, F (202) 380-9129
E-mail: books@american-coatings.com, www.american-coatings.com

Layout: Vincnetz Network, Hannover, Germany

ISBN 3-86630-907-4
ISBN 978-3-86630-907-4

European Coatings Tech Files

Patrick Glöckner

Tunja Jung

Susanne Struck

Katia Studer

Radiation Curing

Coatings and Printing Inks

Technical Basics, Applications and Trouble Shooting

Glöckner/Jung/Struck/Studer: Radiation Curing for Coatings and Printing Inks

ISBN: 978-3-86630-907-4

Foreword

Radiation curing is the fastest growing technology in the inks and coatings industries. It is being widely discussed in numerous publications and conferences and is being used or at least tested in all possible industrial applications.

When asked to write a comprehensive work on radiation curing, we soon realized that the need was not for a theoretical treatise but a book that could be helpful to a large number of readers in their day-to-day work. Our objective was to link theory and practice in a reference guide for formulators in the field of radiation curing. The book is also intended for students and young professionals in coatings and printing inks with no previous experience of radiation curing. In addition to the fundamental principles, therefore, we offer guidance in the selection of raw materials. The Trouble Shooting Chapter summarizes approaches to solving problems that may arise during the manufacture and processing of radiation curable systems.

The focus is on 100 % liquid coating systems that are radically or cationically curable by UV or electron beam radiation. Technologies that are still relatively unimportant are beyond the scope of this book and are discussed only briefly.

The authors thank Evonik Degussa GmbH, Evonik Tego Chemie GmbH, and Ciba Specialty Chemicals for permission to publish this book.

Special thanks go to Kurt Dietliker, Shawn Feeney, Winfried Hamann, Eugene Sitzmann, and Emmanouil Spyrou for their substantial contribution to this work. We are also grateful to Hans-Dieter Christian, Wernfried Heilen, Dirk Hoppe, Claudius Neumann, Friedrich Georg Schmidt, Volker Schrenk, and Thomas Werner for numerous discussions and suggestions.

Patrick Glöckner,
Tunja Jung,
Susanne Struck,
Katia Studer

Essen, Germany and Basel, Switzerland, May 2008

Glöckner/Jung/Struck/Studer: Radiation Curing for Coatings and Printing Inks

ISBN: 978-3-86630-907-4

Content

Glöckner/Jung/Struck/Studer: Radiation Curing for Coatings and Printing Inks

ISBN: 978-3-86630-907-4

1 Introduction

1.1 Definition

Radiation curing coatings (RC) and inks are formulations of components that contain reactive groups which react with each other after exposure to energy-rich radiation. In general there is no physical drying involved in the curing mechanism. Up-to-date formulations tend to be free of organic solvents.

Even though many kinds of radiation would theoretically be possible the industrial application is mainly limited to the cure by ultraviolet light (UV) and electron beam (EB). While in the case of EB the irradiation itself initiates the crosslinking in the case of the UV cure the radiation does not directly activate the reactive groups, but activates a light sensitive component called photoinitiator that actually starts the polymerization reaction. The cure by radiation takes place at very high speed – from parts of a second to just a few seconds.

1.2 Main differences compared to conventional coatings

Conventional coatings consist of

- binders (film forming polymer),
- volatile, organic solvents,
- pigments and fillers (if not a clear coat), and
- additives.

Conventional coatings dry by evaporation of the solvents and possibly a subsequent chemical crosslinking of the binder with itself or a hardener. This may take place at room temperature, at elevated temperatures below 100 °C (enforced drying) or above 100 °C (stoving/baking). If no chemical reaction takes place after the evaporation of the solvent the coating is called physically drying.

The most well-known representatives are solvent-based formulations. Similar principles apply to waterborne coatings where the organic solvent is replaced by water. Here, the binders are often dispersed in water, rather than solved. In any case the solvent is primarily used to allow the manufacturing and the application of the coatings and acts as viscosity reducer. After application it leaves the film completely.

In constrast **radiation curing formulations** cure without any evaporation of solvent. Indeed normally there is no volatile solvent present at all. All components in a radiation curing formulation stay within the film. The film forming agents present are a combination of higher molecular weight oligomers and lower molecular weight monomers, all of which carry reactive groups to crosslink with each other. This means, that none of these is solely a visco-reducer, but all components become part of the final polymer network and contribute to the final film properties. Figure 1 illustrates the differences between conventional and radiation curing coatings.

Glöckner/Jung/Struck/Studer: Radiation Curing for Coatings and Printing Inks

ISBN: 978-3-86630-907-4

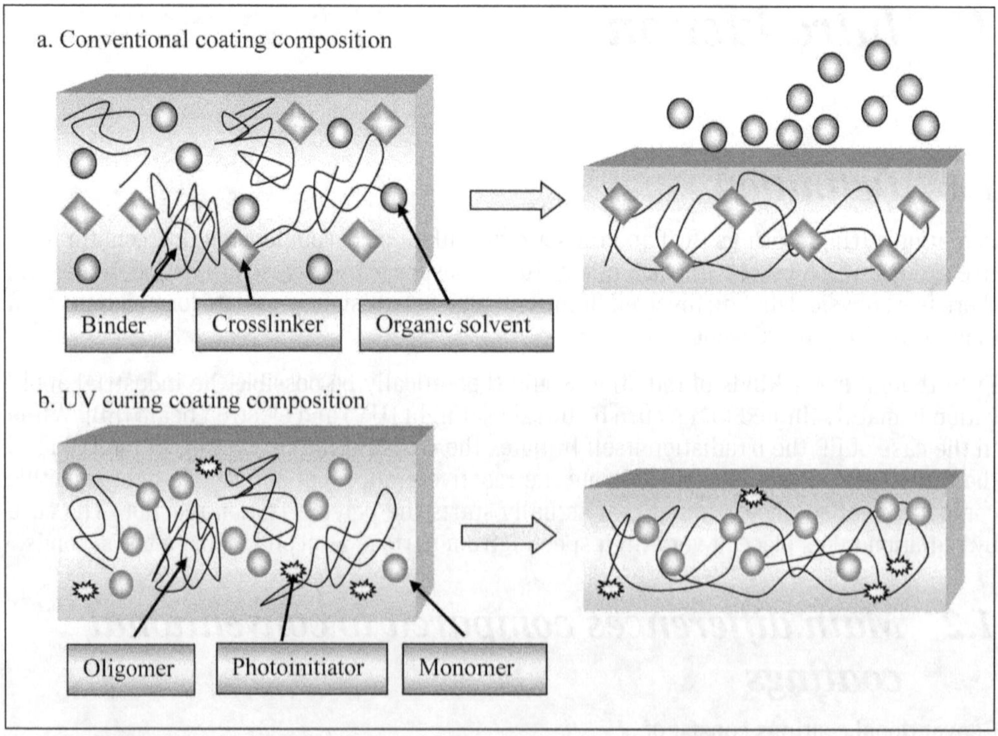

Figure 1: a) In case of conventional coatings organic solvents leave the film during drying and crosslinking; b) UV curing formulations crosslink without emitting any organic solvents

Table 1: Some properties of conventional and radiation curing coatings and inks

	Conventional coating	Radiation curing coating
physical drying	yes	no
chemical crosslinking	yes (depending on binder)	yes
volatile solvents	yes	no
drying times	minutes to days	parts of a second

Radiation curing coatings consist of
- oligomers,
- monomers,
- photoinitiators (in case of UV curing),
- pigments and fillers (if not a clear coat), and
- additives.

In Table 1 some properties of conventional and radiation curing coatings and inks are listed and compared. A detailed explanation of the formulation components can be found in Chapter 3 of this book.

1.3 Advantages and disadvantages

Radiation curing is distinctly different to conventional coating technologies. The nearly immediate cure and the absence of solvents allow obtaining a dry and fully cured film in just a few seconds. There is no need for long flash-off times or time- and energy-consuming drying tunnels in application lines. Cured surfaces can immediately be processed. Printed goods can be coated, cut, embossed, etc. right away; coated surfaces can be sanded, recoated, glued and assembled without waiting time. Radiation curing is the ideal technology for curing coatings on flat (two-dimensional) substrates. Especially when the substrate is fed from a web like in graphic arts applications or flat substrates moved on a conveyor belt through the application line like in wood coatings applications a radiation curing unit is easy to install and operate. Figure 2 shows the principle of a web-fed UV coating unit. The most important advantages and disadvantages of radiation curing over conventional coating technology are summarized in Table 2.

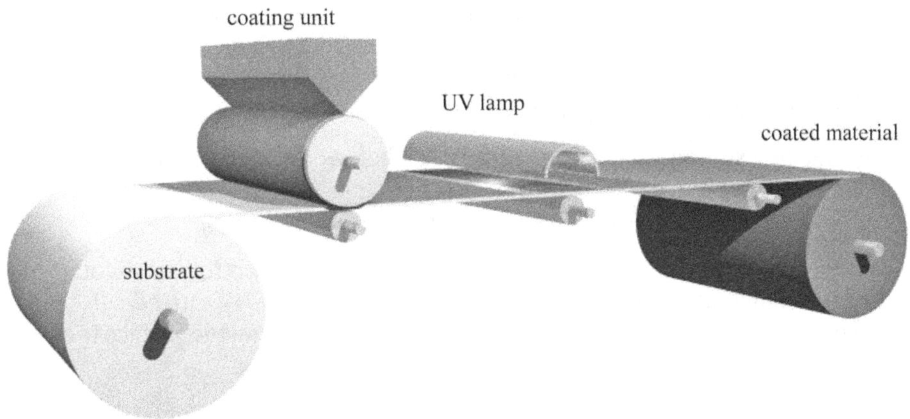

Figure 2: Principle of a web-fed UV coating unit, source: EVONIK TEGO

Table 2: Main advantages and disadvantages of radiation curing over conventional coating technology

Advantages	Disadvantages
very fast cure speed	cure of three-dimensional objectives requires complex curing units (shadow areas)
time and cost savings	skin irritation in case of some raw materials
absence of volatile solvents (VOC), zero to low emissions of VOC	very matte finishes difficult to obtain
cured films display outstanding resistance properties and gloss	high pigmentation levels at high film thicknesses can create cure problems
reduced energy consumption for cure (environment and cost)	inherent viscosity of radiation curing formulations makes certain application methods impossible at room temperature
relatively small capital investment in case of UV unit	running and maintaining the curing unit requires qualified personnel
little space requirements (especially in case of UV units)	partly high raw material costs
application on heat-sensitive substrates possible	relatively high investment costs in case of EB
no pot-life = coating/ink does not dry before irradiation (no problems in case of machine stop; easy to clean and easy recycling)	adhesion especially to metal or some plastic substrates
	odor of formulations

Table 3: Main differences between UV and EB curing

	UV	EB
curing possible in the presence of oxygen	+	-
little capital investment necessary	+	-
fast cure	+	++
low content of extractables	o	+
little space required for curing unit	++	-
easy curing of highly pigmented formulations	-	++

1.4 EB versus UV Curing

So far we have been discussing radiation curing in general, summarizing UV and EB technology. However there are distinct differences between these two which will be discussed further in detail in Chapter 2.3 (Table 5). The main difference from a formulator's perspective is the absence of photoinitiator in EB curing formulations. The high energy level of electron beam radiation initiates polymerization directly.

From a technology standpoint the main difference between the two processes is the kind of radiation. For the UV cure very energy-rich radiation with a wavelength from approx. 180 nm (UV-C) up to 380 nm close to visible light is used.

In EB cure electrons are shot onto the coating by the use of an electron accelerator. Oxygen needs to be excluded efficiently to allow proper cure. Whereas UV curing can take place under atmospheric conditions, EB units need inertization, meaning the surface to cure must be flooded with nitrogen.

The final decision which technology is most suitable for a certain application depends on the specific requirements. As a matter of fact UV curing is much more widely used today than EB curing. In Europe EB technology is used very little overall. But since this technology has its distinct advantages it waits to be seen if this picture will change in the future.

1.5 Historical development

Since the beginning of the 20th century the initiation of a curing process by the exposure to UV light had been known. Already 1946 first patents were filed describing the cure of coatings respectively printing inks by ultraviolet radiation. It was also in 1940s when first wood coatings were cured by irradiation by UV light. First industrial application of electron beam curing coatings took place in the early 1970s. It had taken many experiments and pilot plants until the first commercial electron beam curing unit was installed. It was the Ford Motor Company in the US who started off with EB curing of plastic coatings for interior trim applications [1]. Soon followed furniture coating applications in Germany and the Netherlands. In these early years of radiation curing there was a strong focus on electron beam curing. Numerous patents published in the late 60s in 20th century apply this technology to different applications. In the 80s and 90s UV curing appeared to be the technology of choice. Except for North America EB curing was hardly present in industrial applications.

In any case both technologies have proven successful on a long-term basis in all applications where flat substrates were involved and fast cure speed offered major advantages.

1.6 Market figures and main applications

Accurate global figures are difficult to obtain. But scanning through various market studies and summarizing the findings the following picture is created. Over the last decade radiation curing coatings and inks have enjoyed a global growth of 6 to 10 % every year. The estimated global market size for radiation curing formulations is currently (2007) exceeding 300,000 metric tonnes on a global basis.

Globally the biggest applications for radiation curing formulations appear to be the

- wood coatings
- overprint varnishes (OPV)
- pigmented printing inks
- other industrial coatings (incl. automotive coatings)
- electronics and telecommunication
- adhesives

A rough estimation utilizing figures from various sources indicate the following global volume distributions:

According to an SRI Study from 2005 Asia has the largest consumption of radiation curable materials, closely followed by Europe and North America [2]. Other studies still claim Asia to have a lower consumption than Europe and America. But anyhow Asia is expected to draw at least level very soon.

The growth perspective is excellent. Depending on the sources a further global growth of 6 to 8 % annually is forecasted for radiation curing formulations. China and South East Asia are expected to grow over-proportionally with double digit numbers, North America following with 7 to 8 % per year and Europe and Japan falling behind with 5 respectively 4 % per year.

Even though the volume distribution by region currently appears relatively homogeneous the distribution of these end uses for radiation curing is not. Whereas the usage of RC formulati-

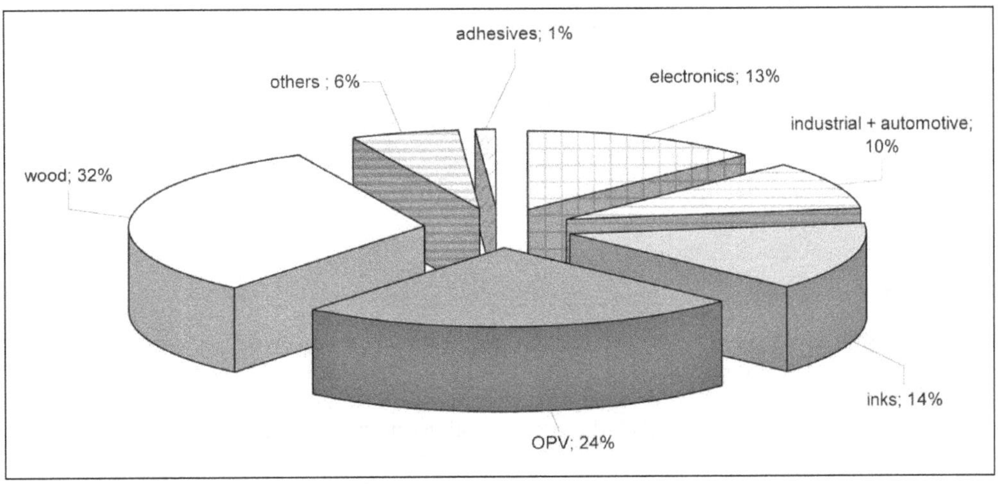

Figure 3a: Estimated volume share of applications for RC formulations

Table 4: Consumption of radiation curable materials by major region [2]

	2001 1,000 metric tonnes	2004 1,000 metric tonnes	2009 1,000 metric tonnes
Europe	84	102	129
North America	68	79	118
Asia (incl. Japan and China)	unknown	120	167
Japan	42	45	56
China	unknown	33	53

ons in Europe is strongly dominated by wood coatings, the graphic arts applications are much more dominant in the NAFTA. Also the applications of plastic coatings, release coatings, optical fibre coatings, and vinyl flooring have a higher share than in Europe. In Japan, still the largest producer of RC formulations in Asia [3], the share of opto-electronic coatings is significantly higher than indicated in the graph above. Details about these applications will be discussed in Chapter 5 of this book.

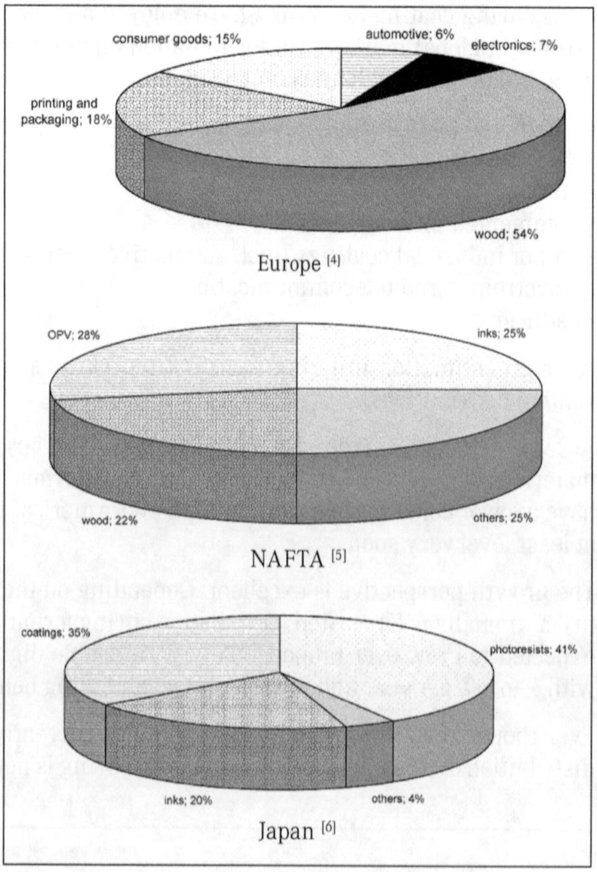

Figure 3b: Split of applications for RC formulations in 2005/2006

By far the largest share of the current 300,000 tonnes of the total radiation curable formulations are cured by UV light. EB technology has only captured 10 to 12 % of the market with the main applications in adhesive laminations, high-gloss varnishes, silicone release coatings, decorative furniture foils and papers, and low odor coatings for flexible food packaging materials.

Cationic curing systems serve mostly specialized niche markets including can and coil coating due to their superior adhesion to metal and sterilizable packaging materials. Their market share is estimated arround 1 and 2 % by volume.

References

[1] P. Garratt, Hrsg. Ulrich Zorll, Strahlenhärtung, Vincentz Verlag, Hannover, Germany, 1996
[2] Summary of SRI study Radiation Curable Coatings, August 2005
[3] Frost and Sullivan, Asian Radiation Cured Formulations Market, Study 2007, Chapter 3.2
[4] Frost and Sullivan, European Markets for Radiation Curable Coatings, Study 2006, Figure 3.5
[5] G. Cohen, Status of UV&EB in North America, Radtech Status Report 8_06,
[6] Frost and Sullivan, Asian Radiation Cured Formulations Market, Study 2007, Chart 7.1

2 Basics of radiation curing technology

Radiation curing is an ultrafast and efficient method to convert a liquid film into a solid material by exposure to radiation [1, 2]. This technology developed in the 1960's gained progressively importance and reached new markets to finally become state-of-the-art in many applications including the coating and ink industry as well as electronic applications. Key features responsible for this growth are the availability of 100 % liquid systems with no VOC and low energy consumption, which are both distinct ecological and economical advantages over thermal or physical drying processes. Additional unique benefits include in the case of radical polymerization an ultrafast process [3, 4], reduced space requirements in a production line, an easy control of the reaction and a curing step performed at "cool" temperature (room temperature) which allows this technology to be applied for applications involving heat-sensitive substrates. Use of light further opens the possibility of image-wise curing, which is at the base of all imaging applications such as printing plate fabrication or electronic applications including chip production, electronic materials or the fabrication of flat panel displays. Final materials cured by this way are strongly crosslinked and exhibit outstanding properties as well as excellent physical and chemical resistances.

Two radiation curing technologies are currently employed in the coating and ink industry: UV curing and electron-beam (EB) curing. Curing induced by UV light currently represents about 90 % of the radiation-curing market in Europe. UV curing, in other terms photoinitiated polymerization, photopolymerization or photocrosslinking, is a polymerization reaction involving a functionalized oligomer (Figure 4) leading to the formation of a solid crosslinked film [5]. The cure mechanism is initiated by a light sensitive molecule called a photoinitiator and can be a cationic [6] or a free radical polymerization reaction [7] as well as a catalyzed polyaddition. Light sources which activate the process are usually medium pressure mercury lamps, xenon lamps, lasers or electrodless vapor lamps emitting mostly between 200 and 500 nm. Near-visible light sources (e.g. sunlight, fluorescent lamp, LED) are currently used for niche applications but are raising interest because of their low toxicity.

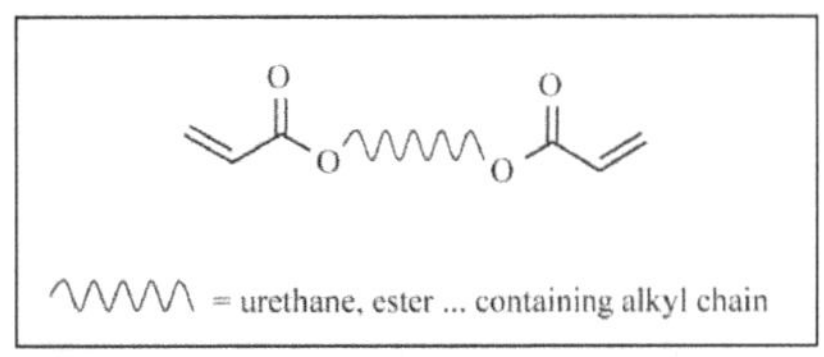

Figure 4: Example of a functionalized monomer contained in a radiation-curable formulation cured by a radical polymerization

Electron-beam curing involves an electron beam which directly induces the curing process of a functionalized oligomer [8] by a free radical mechanism in absence of a photoinitiator. In the case of cationic polymerization induced by electron beam, an initiator is required.

In this chapter, an overview of the UV curing technology will be outlined first, together with the different curing mechanisms, their advantages and limitations. The influence of the cure parameters on the cure speed and the material performances will then be discussed. Basic principles of EB curing will be reviewed in a second part.

Glöckner/Jung/Struck/Studer: Radiation Curing for Coatings and Printing Inks

ISBN: 978-3-86630-907-4

2.1 UV curing technology

Light is a key element in that it initiates the cure process: UV radiation is absorbed by the photoinitiator which undergoes chemical modification leading to the formation of active species which initiate the curing process. Depending on the polymerization mechanism, UV curing can be classified into two major categories which require different types of photoinitiators: free radicals are generated in the case of **free radical UV curing** and cations are produced during a **cationic UV curing** process. Besides these two major types of curing mechanisms, polyaddition or polycondensation reactions are a third type of crosslinking mechanism that can be initiated by light. For such systems, photolatent catalysts that generate an active species – usually an acid or a base – capable of accelerating the crosslinking reaction are required: this cure mechanism is still marginal, but expected to significantly grow within the next few years.

Each of these mechanisms shows distinct advantages and limitations so that they are not competitive but complementary. The composition of free radical or cationic UV curable formulations differs both in the nature of the oligomers and photoinitiators. As a result of the different cure mechanisms, several reactivity parameters are different, while attractive features such as the rapidity of the process (from a fraction of a second for free radical to minutes for cationic) are preserved. As described in more details in Chapter 3, these formulations basically contain the following components:

- **photoinitiator**. A radical photoinitiator produces free radicals under the action of light, while a cationic photoinitiator usually generates protons or a Lewis acid. Figure 5 shows an example of both photoinitiator types. Typical photoinitiator concentrations are ranging from 0.5 to 8 wt% depending on the application.
- **functionalized oligomer** or **binder** bearing the polymerizable groups and constituting the largest part of the formulation, ranging typically from 50 to 80 %. This component builds the film backbone during crosslinking and confers the final mechanical properties and weathering resistance to the material. The type of reactive groups differs whether the curing involves a free radical cure (acrylates, unsaturated polyester/styrene resins, and thiol-enes) or a cationic mechanism (epoxies and vinyl ethers). Figure 6 shows some

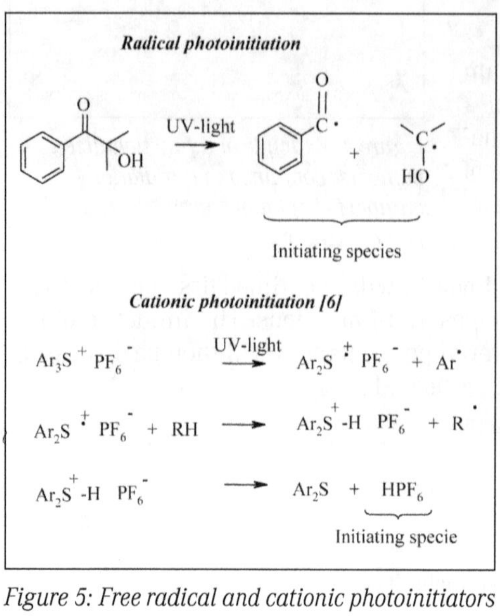

Figure 5: Free radical and cationic photoinitiators and their activation reaction

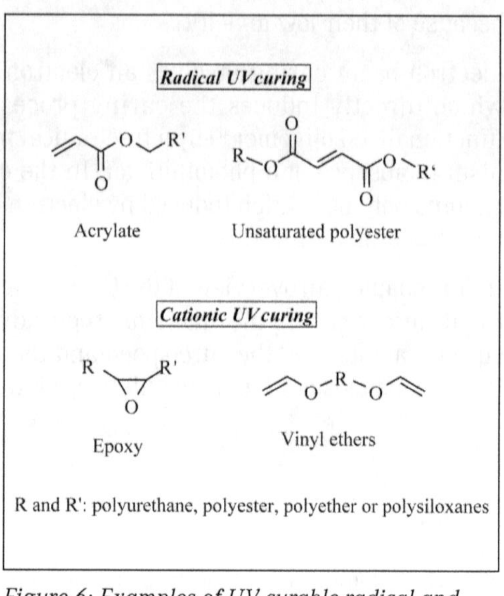

Figure 6: Examples of UV-curable radical and cationic oligomers

of these oligomers: the influence of the reactive groups, the functionality and the backbone on the cure speed and material properties will be detailed in Chapter 3.3.

- a mono- or multi-functionalized **monomer as reactive diluent** to adjust the formulation viscosity (0 to 50% of the formulation). This component copolymerizes with the binder and is thus incorporated into the polymer matrix during the curing process.
- **additives**, such as light stabilizers, leveling agents, pigments, which either facilitate the application step or confer some properties such as durability or color to the final material (see Chapter 3.5).

Similarly to a standard bulk polymerization, a UV curing process has an initiation, a propagation and a termination step as represented in Figure 7. In the case of a photopolymerization, the initiation step is triggered by light. Propagation, which converts the liquid monomers and oligomers into a solid polymer network, and termination, deactivating most of growing radicals or cationic chains, are thermal chain reactions which follow the same kinetics as conventional polymerization reactions. In reality bulk photopolymerization is a much more complex process where the progressive viscosity increase reduces the chain mobility thus affecting the polymerization kinetics (see Chapter 3.3.1).

2.1.1 Free radical UV curing

The rapidity of free-radical UV curing is one major advantage responsible for the constant growth of this technology: curing proceeds within a fraction of a second. This ultrafast transformation from a liquid phase to a solid state allows increasing productivity on a production line while maintaining energy cost at a low level.

2.1.1.1 UV exposure and initiation

UV-light initiates the cure process by activating the photoinitiator to generate free radicals. This step, called photolysis, is represented in Figure 8a. These radicals subsequently add to the double bond of oligomers or monomers, thereby forming a carbon centered radical in the initiation step of a radical polymerization reaction (Figure 8b).

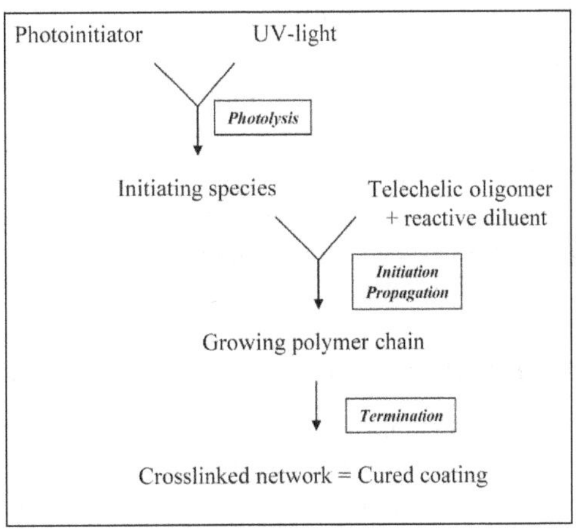

Figure 7: Principle of a UV curing process

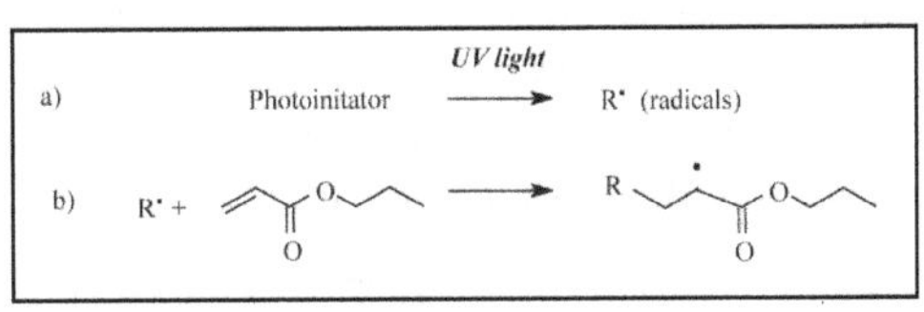

Figure 8: a) Photolysis and b) cure initiation

To make the initiation possible, the photoinitiator has to be capable of absorbing light generated by the lamp, i.e. the absorption spectrum of the photoinitiator has to overlap with the emission spectrum of the lamp as shown in Figure 9. In the case that the photoinitiator does not meet this fundamental requirement, no curing reaction is initiated and films remain liquid. The useful absorption of commercial photoinitiators usually extends from UV-C around 230 nm up to UV-A in the range of 380 to 400 nm. For specific applications, some photoinitiators absorb slightly above 400 nm, while photoinitiators absorbing strongly in the visible range are employed for niche applications (see Chapter 3.2).

Medium pressure mercury lamps, emitting between 220 and 580 nm are well adapted

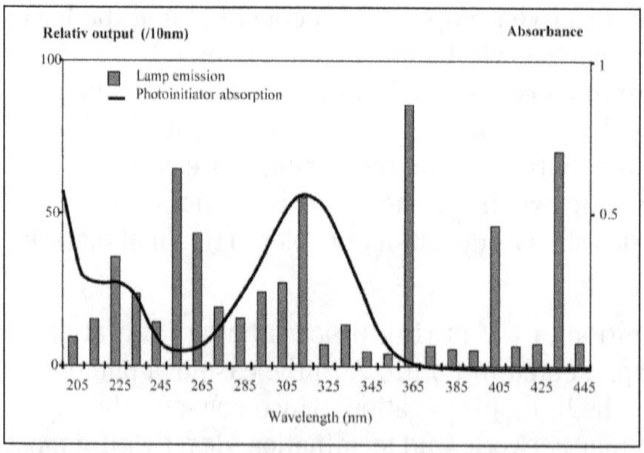

Figure 9: Absorption spectrum of a photoinitiator and emission spectrum of a medium pressure mercury lamp

for this purpose. For safety concerns, new lamps emitting mostly in the UV-A range have been recently introduced into the UV curing market (see Chapter 4).

A number of other parameters play a major role in influencing the initiation rate and therefore the cure efficiency of free-radical UV curing:

- quantum yield of the photoinitiator, i.e. a parameter characterizing the amount of initiating species generated per photon absorbed, which has to be as close as possible to the theoretically possible value
- an increase of the photoinitiator concentration also facilitates the cure process under most conditions [9]
- high light intensity improves the cure speed, as an important amount of initiating species are generated simultaneously. However, above a certain limit, increasing the light intensity does not affect the cure speed anymore [10]
- in the case of photoinitiated radical polymerization, if photoinitiator concentration and light intensity are low, oxygen inhibition affects the initiation efficiency and the polymerization process. Performing the curing under an inert atmosphere addresses this issue.

The dependency of the initiation efficiency with respect to these parameters is well understood and can be quantified by the equation relating the rate of initiation to the light intensity, photoinitiator type and concentration [11, 12]:

$R_I = \Phi_I I_0 [1-\exp(-2.3 A)]/l$, with

R_I = rate of initiation [$mol \cdot L^{-1} \cdot s^{-1}$]
Φ_I = quantum yield of initiating species [$mol \cdot E^{-1}$], which depends on the photoinitiator
I_0 = intensity of incident light [$E \cdot s^{-1} \cdot cm^{-2}$]
$A = \log(I_0 / I_t) = \varepsilon\, l\, [PI]$ = intrinsic absorbance
l = film thickness [cm]
I_t = intensity of transmitted light [$E \cdot s^{-1} \cdot cm^{-2}$]
ε = molar extinction coefficient of the photoinitiator [$L \cdot mol^{-1} \cdot cm^{-1}$]
[PI] = photoinitiator concentration [$mol \cdot L^{-1}$].

2.1.1.2 Propagation – curing mechanism

As represented in Figure 10 for an acrylate polymerization, radicals formed in the initiating step (Figure 8b) further propagate by a free radical mechanism. The propagation rate depends on the initiation rate as well as on the oligomer reactivity, but viscosity progressively increases thus slowing down the polymerization kinetics. The propagation rate is little influenced by temperature, i.e. a temperature increase usually does not affect the reactivity of the system. However, a high temperature increases the chain mobility during gelation, thus allowing a faster and more extensive cure [9]. Local temperature increase can be reached under intense illumination by the heat generated by the lamp and the exothermicity of the reaction (see Chapter 3.3.1).

Because of excellent performances in terms of cure speed and final properties (weathering resistance, mechanical properties), acrylate functional oligomers and monomers have raised a lot of interest and now occupy a predominant position in the UV curing market. A second category of binders having a significant but decreasing importance are low cost unsaturated polyester/styrene resins (Figure 11) which are still widely used in wood applications in spite of many disadvantages, namely:

- a low cure speed
- inferior properties compared to acrylates
- the volatility of the components (styrene)
- high sensitivity to oxygen inhibition.

Thiol-ene formulations are a third type of free radical UV curable systems [13] and will be described in Chapter 3.3.2.7.

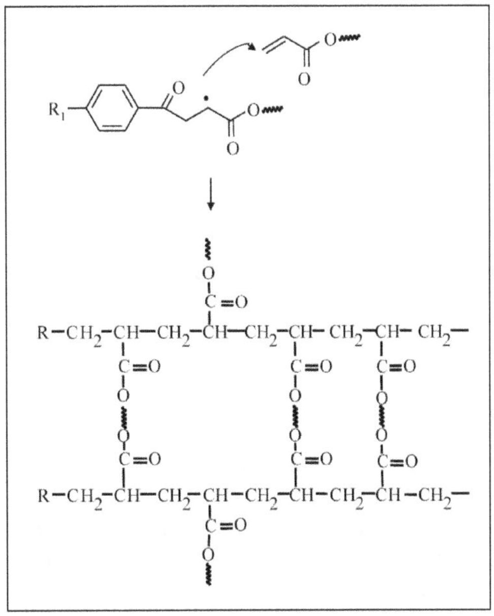

Figure 10: Radical curing of an acrylate resin

2.1.1.3 *Chain termination*

Reaction diffusion controls the termination process: radicals either undergo a recombination reaction ($P_n\cdot + P_m\cdot \rightarrow P_{n+m}$) or a disproportionation ($P\text{-}CH_2\text{-}CH_2\cdot$ + $P\text{-}CH_2\text{-}CH_2\cdot \rightarrow$ $P\text{-}CH_2\text{-}CH_3$ + $P\text{-}CH{=}CH_2$), or remain trapped due to the vitrification of the highly photocrosslinked network, what also stops prematurely the polymerization and frequently leaves unreacted monomer molecules.

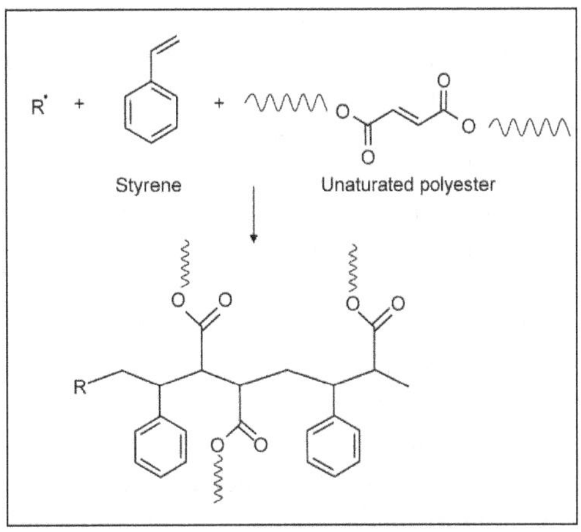

Figure 11: Curing of a polyester/styrene resin

2.1.1.4 *Unique features*

Major benefits of free radical UV curing are:

- ultra-rapid hardening, as the process occurs within a fraction of a second at room temperature
- low energy consumption which makes UV curing a green and attractive technology
- "cold curing": ideally adapted to coat heat-sensitive substrates
- reduced space requirement of a UV-production line in comparison to ovens required for thermosetting cure
- reaction is easily controlled: light initiates the cure process in exposed areas allowing a spatial and temporal control of the reaction. The polymerization stops as soon as light is removed
- no volatile solvents are required to adjust the viscosity. Reactive diluents used to this end are incorporated in the polymer matrix (low VOC)
- formulations are stable for an extended time in the absence of UV-light (usually several months or longer) and do not have a potlife
- production of tailor-made coatings: wide choice of oligomers, reactive diluents and photoinitiators and easy accessibility to a broad variety of commercially available raw material.

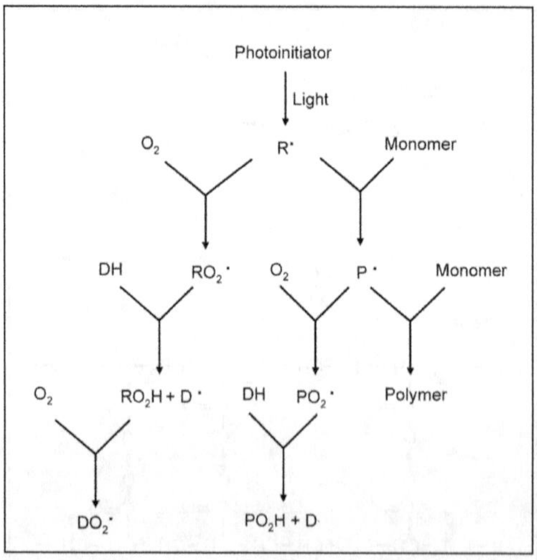

Figure 12: Oxygen inhibition during free radical UV curing; P = Polymer, DH = Hydrogen Donor, R = Radical

Limitations of free radical UV curing are:

- similarly to all radical reactions, oxygen inhibits free radical UV curing to some extent (Figure 12). Oxygen inhibition is particularly pronounced when irradiating conditions generate a low amount of radicals, i.e. for short exposure at a low light intensity in the presence of small quantities of photoinitiator (see Chapter 4). The effect is also marked in the curing of thin layers where diffusion of oxygen into the formulation is very efficient. This unwanted reaction between propagating radicals and oxygen leads to tacky surfaces or to materials exhibiting poor performances. Curing under inert atmosphere (e.g., "Larolux" [14]) is one way to overcome this issue.
- so-called inner filter effect of the UV curable resin is responsible for a cure gradient from the surface to the depth of the film: oligomers, photoinitiators and photoproducts absorb UV-light thus reducing the amount of light available for the cure initiation in deeper layers. Specific photoinitiators (e.g. acylphosphine oxides) reduce this effect and allow one to cure clearcoats up to several millimeters or even bulk materials of some centimeters thickness.
- low adhesion on non swelling substrates like metal or some plastics (see Chapter 3.3)
- volume shrinkage of acrylates during polymerization is a major concern for curing on flexible surfaces and also affects the film/substrate adhesion [15] (see Chapter 3.3.1.4.1)
- some fillers act as light absorbers or reflectors (e.g., pigments) and prevent the light from reaching deep-lying layers thus limiting the cured thickness to at most 100 µm with some especially designed photoinitiators (see Chapter 3.2)
- UV curing on 3-dimensional substrates has been a major concern for the last years as films remain liquid in shadow areas: new developments, such as special lamps and robots or new UV curing technologies, e.g. "PlasmaCure" provide well-adapted solutions for UV curing in shadow areas. However, they are at an early stage of developments and cannot be applied at an industrial stage so far (see Chapter 4.1.2.4).

2.1.2 Cationic UV curing

Cationic UV curing involves the photogeneration of cations – usually strong *Brönsted* or *Lewis* acids – which are capable of initiating a cationic polymerization. Major advantages of cationic UV curing over a free-radical process are the insensitivity of the reaction towards oxygen inhibition and a lower shrinkage due to the ring-opening polymerization step. Main applications for this technology can be found in printing inks, overprint varnishes and composites, requiring either high production rates under air or specific mechanical properties.

2.1.2.1 Initiation – UV exposure

Cationic photoinitiators are commonly based on sulfonium ($Ar_3S^+X^-$) or iodonium salts ($Ar_2I^+X^-$): the initiation reaction is illustrated in Figure 13.

Light is required to activate the photoinitiator (photolysis a)), while initiation b) and propagation further proceed in the dark as long as the reaction is not completed and the acid is not consumed by impurities. This is in sharp contrast to free radical polymerization which comes to an end when the light is switched off due to the fast consumption of the radical species by recombination or other terminating reactions.

a) $Ar_2I^+X^- + DH \rightarrow ArI^{+\cdot}X^- + D\cdot + HX + Ar\cdot$ (photolysis of the photoinitiator)

b) M (Monomer) + $HX \rightarrow HM^+X^-$ (cure initiation)

Figure 13: Photolysis and initiation of a cationic UV-curing, with Ar = Aromatic Ring , X = Counterion

To allow a fast and efficient polymerization, nucleophilicity of counterion X has to be as low as possible: SbF_6^-, PF_6^-, AsF_6^- are useful counterions for cationic photoinitiators. The different categories of existing cationic photoinitiators will be developed in Chapter 3.2.3. According to the mechanism shown in Figure 13, cations and radicals are simultaneously generated during the photolysis: this phenomenon raised some interest a few years ago for hybrid radical and cationic systems [16], but has found a limited use in commercial applications. Similarly to free radical UV curing, absorption spectrum of the photoinitiator, emission of the lamp, light intensity, photoinitiator reactivity and concentration play a key role in controlling the initiation step.

2.1.2.2 Propagation – curing mechanism

Instead of involving the addition of a radical to a double bond, cationic UV curing results from the attack of the proton on an electron rich group such as vinyl ethers or epoxies, as shown in Figure 14 and Figure 15.

Unlike radicals which recombine or undergo disproportionation reactions resulting in the consumption of the radical species, cations continue to propagate a long time after irradiation (so-called post-cure) due to the living character of cationic polymerization. Theoretically this leads to a complete consumption of the oligomer: in reality, impurities (nucleophilic species such as water or bases) prematurely inhibit the reaction.

Propagation rate constants are lower for epoxy than for acrylate polymerization and, depending on the type of epoxies, result in cure times of a few seconds or up to several hours at room temperature [17]: post-cure can be accelerated by heating the sample, as cationic polymerizations are thermally controlled.

Epoxies occupy the largest place in the cationic UV curing market due to a strong interest also for thermal curing and a relatively easy accessibility to a wide choice of products. Vinyl ethers, which are usually more reactive than epoxides (Figure 16 [18]) are of second commercial importance.

Other heterocycles polymerize by cationic ring opening, e.g. oxetanes, lactones. Different categories of oligomers available on the market will be developed in Chapter 3.3.4. *Lewis*

$$HM^+X^- + nM \rightarrow H(M)nM^+X^-$$

Figure 14: General mechanism of cationic polymerization M = Monomer, X = Counterion

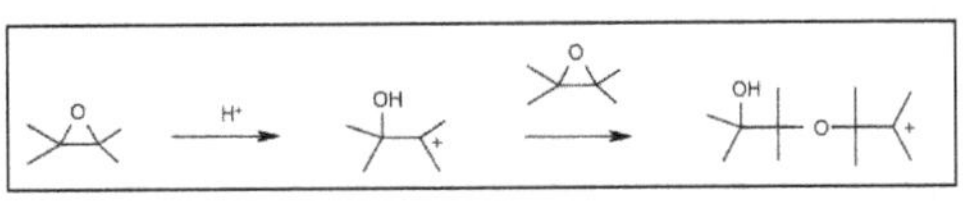

Figure 15: Cationic polymerization of epoxies

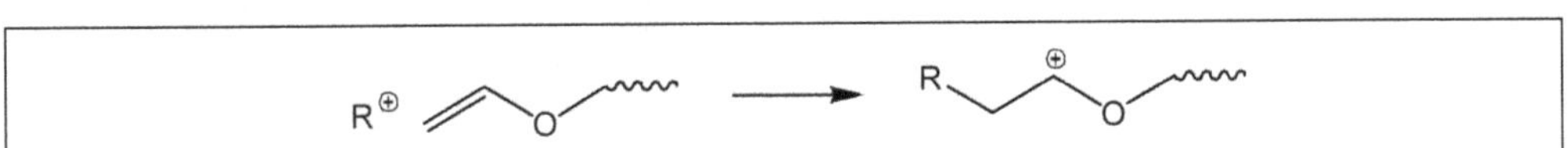

Figure 16: Cationic polymerization of vinyl ethers

or *Brönsted* acids are also capable of catalyzing other type of crosslinking reactions such as polycondensation or polyaddition reactions of acetales or siloxanes. While photolatent strong acids are required for epoxy or vinyl ether polymerization, many of the other crosslinking reactions are also catalyzed by weaker acids, allowing e.g. the use of neutral latent sulfonic acid derivatives.

2.1.2.3 Unique features

In spite of a smaller commercial significance compared to free radical UV curing, cationic photopolymerization of epoxies has found a number of applications due to unique advantages inherent to the epoxy chemistry, among them:

- shrinkage is less pronounced than with acrylates resulting in an improved adhesion
- outstanding coating properties, such as high chemical resistance, high gloss, high mechanical strength and abrasion resistance
- high thermal resistance
- post-cure after UV-irradiation
- no oxygen inhibition.

Limitations of cationic UV curing are:

- water quenches the polymerization reaction: cure is therefore strongly inhibited by moisture and more generally by all bases (alcohols, hydroxyl containing compounds) acting as chain transfer agents. These impurities cause premature polymerization termination, what is a major concern when ambient humidity is high, adversely affecting the performances of the material
- curing is much slower compared to a free radical process but continues during a certain period of time after the end of the irradiation
- presence of acidic species in the final material is likely to promote corrosion
- similarly to radical UV curing, light is required to initiate the process: shadow areas and filled systems therefore limit the application field of cationic UV curing. This major concern is addressed by developing new photoinitiators and new technologies
- also in the case of a cationic photopolymerization, the inner filter effect is responsible for a cure gradient from the surface to the depth of the film
- availability of raw materials is reduced compared to acrylates

Recent developments in the UV curing area raise new interest as they address most of those issues, broaden the application field of radiation cure technologies [19,20,21,22] and open up many new market opportunities.

2.1.3 UV curing by photolatent bases

A new generation of UV curable systems was recently introduced in the coating industry: photolatent catalysts allow a light-controlled activation of a catalyst for the curing of coating systems requiring such accelerators, usually a base. By this way, fairly stable formulations (e.g. 2-pack polyurethanes) can rapidly cure at a relatively low temperature once applied onto a substrate and exposed to light. This new technology should simultaneously address oxygen inhibition in UV curing and shadow areas, as a slow cure still

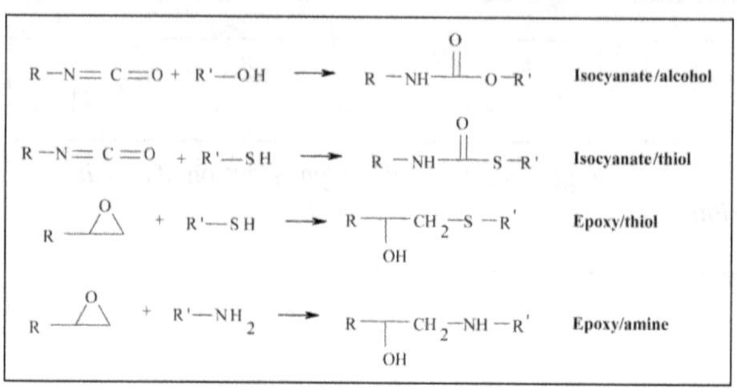

Figure 17: Example of some industrial 2-pack coating systems

proceeds even in the absence of light. These products have been especially designed for 2-pack systems frequently employed in the coating industry as represented in Figure 17.

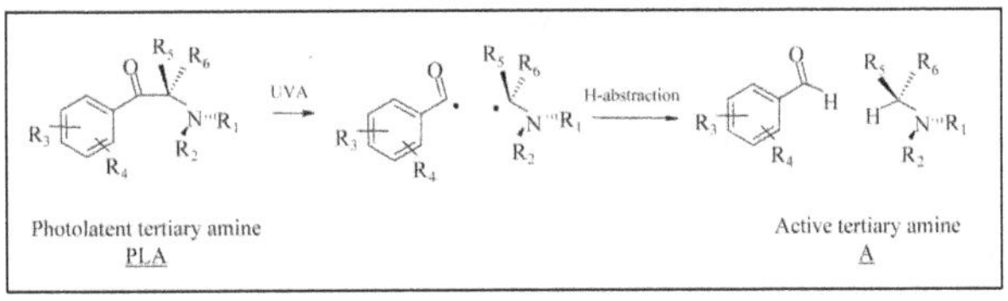

Figure 18: Activation of tertiary amine PLA by UV exposure

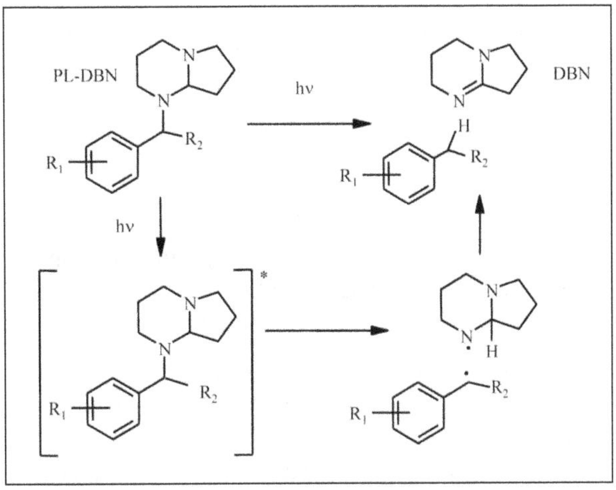

Figure 19: Light activation of the photolatent DBN (1,5-Diaza-bicyclo [4.3. 0] -non-5-ene)

Conventional curing temperatures for these thermal curable coating systems are ranging from 80 to 140 °C, sometimes even higher. This requirement makes these coatings unsuitable for applications involving heat-sensitive substrates. Catalysts such as metal organic compounds or amines are often added to the formulation to rapidly cure at low temperature and give tack-free films within a few minutes at the expense of the formulation stability. Photolatent catalysts provide a fairly stable system that cures rapidly at room temperature after light activation. These products being currently at an early stage of development, principle and recent updates are quickly summarized thereafter:

- α-amino acetophenone PLA [23] releases an active tertiary amine after the action of light, as shown in Figure 18. Active amine A exhibits a higher pKa value than PLA and is sterically less hindered: curing between isocyanate and alcohol, thiol or epoxy and carboxy systems is effectively accelerated by A while the latent molecule PLA does not affect the stability of the formulation in the dark.
- photolatent DBN (1, 5-Diaza-bicyclo [4.3.0] -non-5-ene, PL-DBN) has also been developed as latent catalyst for 2-pack polyurethane, epoxy/thiol or Michael systems. The extreme activity of the non-nucleophilic base DBN restricted its use in the coating industry so far. Now, the **in-situ** activation of this catalyst by UV-light represented in Figure 19 should open up many new opportunities as it allows unproblematic formulating of the coating system and the release of high amounts of catalyst.

Main benefits of photolatent base systems lies in the fact that this new kind of UV curing process is not sensitive to oxygen inhibition and allows a slow shadow cure. In contrast to free radical and cationic photopolymerizations that proceeds via a chain propagation mechanism, the components shown in Figure 17 crosslink via a step-wise polymerization.

2.2 Electron-beam technology (EB curing)

The growing interest for EB curing should not be ignored by the UV curing industry. Both technologies are often regarded as competitive ones but it should be actually more appropriate to consider that they are complementary, as EB curing gives access to some applications which are not accessible using UV curing so far (e.g., curing of thick pigmented sections).

A strong argument against EB curing are the prohibitive cost of the equipment as well as a high energy consumption, which are two major issues EB equipment manufacturers are addressing in their constant concern for an improved accessibility to their equipment.

2.2.1 Principle

In general, EB curing formulations contain nearly the same raw materials as UV curing systems but a free radical photoinitiator is not required. This technology uses an electron beam to cure acrylate-based inks and coatings by a radical mechanism. The irradiation being higher in energy than UV-light, it directly initiates the curing at room temperature thus avoiding the need for a radical initiator. Once electrons collide the coated substrate, they transfer part of their energy to the electrons of the double bond which become excited and initiate the acrylate curing. In addition, the electrons generate radicals at random sites in the polymer backbone, which may result in additional crosslinking via radical-radical recombination or attack of the radical upon an acrylate group. However, depending on the structure of the polymer backbone, fragmentation may also occur and at high doses a reduction of molecular weight is observed. As a consequence the same formulation cured by UV-light and a photoinitiator will not show exactly the same properties [24].

As represented in Figure 20, the electron gun consists of a tungsten cathode heated by an electron flow: electrons are emitted and accelerated under vacuum by a negative voltage produced at the anode thereby generating the electron beam. This beam passes a titanium foil exit window which isolates the vacuum chamber from the surrounding atmosphere and finally bombards the coated substrate under an inert atmosphere [25]. X-rays can be generated when electrons strike the surface, requiring the installation of an additional lead shielding for safety reasons. Typical acceleration voltages required to accelerate the electrons are comprised between 70 and 500 kV. The electron energy (in keV) directly depends on this value.

The **depth of penetration** of the beam into the film directly depends on the electron energy [25] and on the density of the coating or ink formulation. These depths of penetration are usually expressed in g/m^2 and are typically comprised between 20 and 500 g/m^2, which give the depth of penetration in µm in a film having a density of 1. The high energy of the electrons allows penetration into deep layers of thick pigmented films provided the energy of the incident radiation is sufficiently high.

The **dose** applied on the film controls the cure efficiency and depends on the electron current applied on the filament and on the web speed. This quantity is expressed in KiloGray kGy (1 kGy=1000 J/kg) or MegaRads. Typical doses range between 5 and 100 kGy. No initiator being required, the reactivity and structure of the binder determines the reactivity of the formulation, while other parameters such as sample temperature and viscosity are expected to have a significant effect on the cure speed and efficiency.

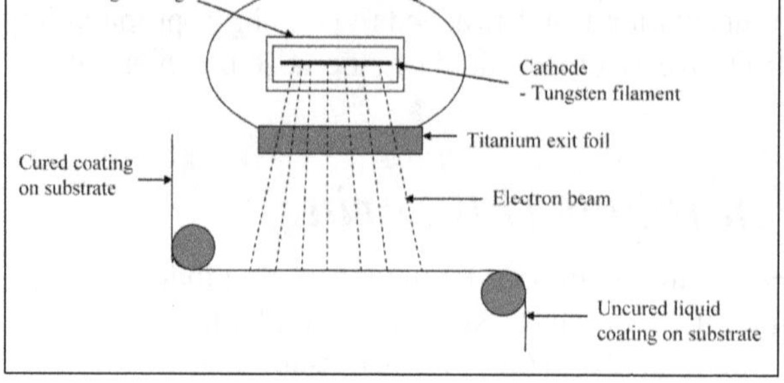

Figure 20: Schematic representation of EB curing equipment

2.2.2 Unique features

The depth of penetration of the beam into the film is affected by the material density: since it is not influenced by a chromophore, an electron beam of sufficient energy can readily penetrate opaque and dense materials which are not transparent to UV light. This mainly justifies the strong success of this technology in applications such as printing inks, adhesives and composites. Attractive features of EB curing can be summarized as following:

- uniform polymerization throughout the film
- high throughput: ultrafast hardening at room temperature, web speed of several hundreds m/min
- no volatile solvents required
- production of tailor-made material
- no photoinitiator required, what eliminates the risk of migration of photoinitiator or of photo-by-products.

Limitations are:

- strong oxygen inhibition in the case of radical EB curing requiring the installation of an inerting system (nitrogen, oxygen content $\ll$200 ppm)
- design of equipment limits the application of such technology to curing on flat 2D substrates
- so far, prohibitive equipment cost and complex installation
- high energy consumption

2.2.3 Electron-beam for cationic curing

EB curing is not limited to applications involving radical curing: also cationic polymerization of epoxies can be achieved using this technology. This process is steadily gaining interest, as it allows an efficient cure of thick pigmented epoxy or vinyl ether resins while minimizing oxygen inhibition. A lot of efforts have been devoted to the understanding of this process: it has been proven that the addition of an onium salt (typical cationic photoinitiator) in cationic-curable formulations is essential. In fact, epoxy groups do not undergo a polymerization reaction upon ionization by electron impact but onium salts are well known to decompose upon reduction via the take off of an electron, thereby producing the acid required for the polymerization of the epoxy binder. Furthermore, also anions such as PF_6^- are sought to undergo decomposition upon interaction with high energy electrons, producing initiating species such HF and PF_5. Thus selection of an appropriate cationic initiator for the EB curing of epoxies allows reducing energy consumption [26,27,28] and/or eliminating blanketing.

2.3 *Summary and comparison of the different radiation curing technologies*

Table 5: Comparison between UV and EB curing

	UV curing free radical	cationic	photolatent bases	EB curing radical	cationic
Photoinitiator required	yes	yes	yes	no	yes
Oligomers of commercial importance	acrylate unsaturated polyesters	epoxies vinyl ethers	two-pack coating systems	acrylate unsaturated polyesters	epoxies vinyl ethers
Initiated by	UV or visible light	UV or visible light	UV or visible light	electron beam	electron beam
Exposure atmosphere	air/inert	air	air	inert	inert/(air)
Moisture inhibition	no	yes	no	no	yes
Cure speed	fraction of a second	seconds to minutes	minutes to hours	fraction of a second	seconds to minutes
Temporal control	yes	post-cure	post-cure	yes	post-cure
Cure homogeneity	gradient	gradient	gradient	mostly homo-geneous	mostly homo-geneous
Pigmented systems	limited	limited	limited	yes	yes
Curing on 3D substrates	yes [29]	yes	yes	no	no
Adhesion on substrate	good	excellent	excellent	good	excellent
Shrinkage	yes	little	little	yes	little
Energy consumption	low	low	low	high	high

References

[1] C. Decker, Pigment & Resin Technology, 30, 2001, p. 278
[2] C. Decker, Macromol. Rapid. Commun., 23, 2002, p. 1067
[3] C. Decker, K. Moussa, Makromol. Chem., 189, 1988, 2381
[4] C. Decker, K. Moussa, Macromolecules, 22, 1989, p. 4455
[5] C. Decker, Polym Int., 51, 2002, p. 1141
[6] J. V. Crivello, Ann. Rev. Mater. Sci., 13, 1983, p. 173
[7] C. Decker, B. Elzaouk, Eur. Polym. J., 3, 1995, p. 1155
[8] P. Holl, PPCJ, Feb, 1998, p. 25
[9] K. Studer, C. Decker, E. Beck, R. Schwalm, Prog. Org. Coat., 48, 2003, p. 92 and 101
[10] L. Lecamp, B. Youssef, C. Bunel, P. Lebaudy, Polymer, 38, 1997, p. 6089
[11] C. Decker in Radiation Curing: Science and Technology, S. P. Pappas ed., Plenum Press, New York, 1992
[12] C. Decker, Polym. Int., 45, 1998, p. 133
[13] C. Hoyle, T. Clark, T. Yeon Lee, T. Roper, B. Pan, H. Wei, H. Zhou, J. Lichtenhan, Technical Conference Proceedings - Radtech Europe Conference, Barcelona, November 2005
[14] E. Beck, Technical Conference Proceedings - Radtech Europe Conference, Basel, November 2001, p. 643
[15] T.H. Chiang, T.E. Hsieh, Int. J. Adhes. Adhes., 26, 2006, p. 520
[16] H.J. Timpe, Pure&Appl. Chem., 60, 1988, p. 1033
[17] C. Decker, K. Moussa, J. Polym. Sci., Part A: Polym. Chem., 28, 1990, p. 3429
[18] C. Decker, C. Bianchi, D. Decker, F. Morel, Prog. Org. Coat., 42, 2001, p. 253
[19] T. Jung, P. Simmendinger, K. Studer, W. Tobisch, Technical Conference Proceedings - UV & EB Technology Expo & Conference, Chicago, April 2006
[20] K. Maag, W. Lenhard, H. Löffles, Prog. Org. Coat., 40, 2000, p. 93

[21] K. Studer, T. Jung, K. Dietliker, J. Benkhoff, E. Sitzmann, N. Dogan, Technical Conference Proceedings - UV & EB Technology Expo & Conference, Chicago, April 2006
[22] E. Beck, PPCJ, 193, 2003, p. 15
[23] K. Dietliker, T. Jung, J. Benkhoff, Technical Conference Proceedings - UV & EB Technology Expo & Conference, Charlotte, May 2004, p. 217
[24] R.S. Davidson in Radiation Curing in Polymer Science and Technology, Volume III, Polymerisation Mechanisms, J.P. Fouassier, J.F. Rabek eds., Elsevier Applied Science, London, 1993, chapter 9
[25] [] P. Holl, PPCJ, Feb, 1998, p. 25
[26] [] J.V. Crivello, Radiation Physics and Chemistry, 63, 2002, p. 21
[27] C. J. Janke, G. F. Dorsey, S. J. Havens, V. J. Lopata, Technical Conference Proceedings – 41st Intl. SAMPE Symposium, March 1996, p. 196
[28] B. Defoort, F. Bourserequ, G. Larnac, H. Degrand, F. Cazauy, X. Coqueret, Technical Conference Proceedings - Radtech Europe Conference, Berlin, November 2003, p. 629
[29] T. Jung, P. Simmendinger, W. Tobisch, Eur. Coat. J., 4, 2005, p. 138

3 Raw materials

3.1 Components of radiation curable coatings and printing inks

The components of radiation curable (RC) coatings and printing inks are in principle similar to those of conventional solvent-based coating systems. However there are some major differences in the raw materials. First, the use of volatile organic solvents is not necessary, which leads to a VOC (volatile organic compounds) content close to zero. Secondly, radically and cationically UV curable formulations contain photoinitiators that initiate chemical crosslinking via polymerization of the film-forming components. RC coating systems comprise the following materials:

Monomers and oligomers

Coatings and inks consist largely of oligomeric or monomeric substances. An important characteristic of these substances is the presence of functional groups that polymerize radically or cationically.

Oligomers are also often called (telechelic) prepolymers, resins, or binders. They are the main components of a formulation, and are therefore responsible for the main properties of the liquid formulation and the cured film. Because volatile solvents are not present, oligomers have low molecular weights compared with the binders in conventional systems, which ensures that the formulations have low viscosities.

Monomers, also termed reactive solvents or diluents, are an alternative to volatile organic solvents as they reduce the viscosity of the formulations during production and application and are further chemically incorporated into the film during light exposure. They therefore have a major impact on the physical properties of the cured film. Because some monomers can have rather high molecular weights, the differences between monomer and oligomer are not sharply defined, but rather a matter of definition.

Photoinitiators and photosynergists (sensitizers)

Photoinitiators are components that form reactive species when irradiated, thus initiating polymerization of the monomers and oligomers in cationically or radically curable coating systems. Photosynergists are used in combination with photoinitiators to support the curing process either by increasing the rate of polymerization or shifting the spectral sensitivity of the formulation. In the special case of radical electron-beam curing, no initiators are required as the high-energy radiation directly initiates the polymerization process.

Additives

Additives are used in small quantities to improve particular properties such as substrate wetting, leveling, slip, and scratch resistance, as well as defoaming and deaeration [1].

Glöckner/Jung/Struck/Studer: Radiation Curing for Coatings and Printing Inks

ISBN: 978-3-86630-907-4

Colorants and fillers

Colorants are responsible mainly for coloration, and are divided into insoluble pigments and soluble dyes. Pigments also perform additional functions such as supporting corrosion protection. The colorless fillers are insoluble in the coating medium. They improve, for example, the rheological properties of liquid formulations or the mechanical properties of the cured film [2].

The choice of raw materials depends upon the expected performance and area of application. The next section covers the typical components of radiation curable coatings and printing inks and suggestions for their use.

3.2 Photoinitiators and photosensitizers in UV curing

3.2.1 Generalities

Photoinitiators are a key element of the formulation making the link between UV light and the cure process. They control the cure speed and efficiency both at the surface and in deep-lying layers to a considerable extent. Photoinitiating systems are based on a molecule or a mixture of molecules absorbing UV light to reach an excited state from which they undergo a photochemical reaction generating the initiating species [3]. A simplified pathway followed by the molecule is represented in Figure 21 which shows the energetic activation of the photoinitiator leading to the initiation of the curing process via a reaction from the triplet state. A more explicit illustration of the photochemical reaction is given in Figure 22 with the example of a typical radical photoinitiator.

To enable a photochemical reaction, the photoinitiator has to contain a group capable of absorbing light: this chromophore is an extended conjugated π-system that can absorb light of useful wavelengths. For photoinitiators absorbing in the UV region, the chromophore often consists of a benzoyl type moiety which may be further substituted in order to shift the absorption into the most useful region. Moreover, the keto group in these systems is well known to be at the base of very efficient photoreactions that can be used for the initiation process.

The efficiency of a photoinitiator is in part controlled by its quantum yield. Quantum yields referring to different events can be used: the **quantum yield of photolysis** ϕ_{ph}, is defined as the ratio between the number of molecules undergoing a photochemical reaction to the number of absorbed photons (maximum value of ϕ_{ph} is one). The **quantum yield of radical formation** ϕ_r [4] is defined as the ratio between the number of initiating radicals formed to the quanta of light absorbed.

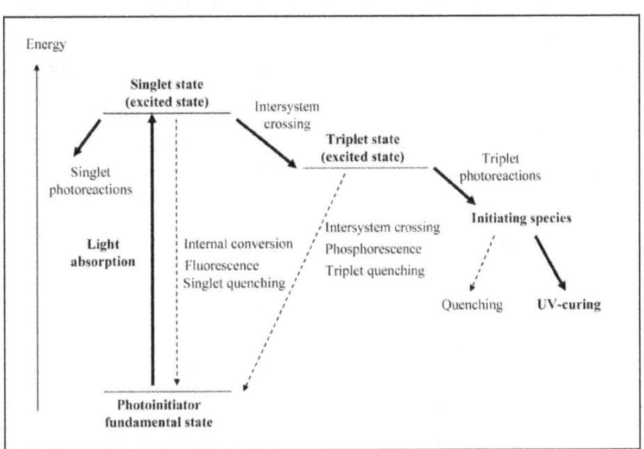

Figure 21: Energy levels of a photoinitiator upon light absorption

Highly efficient photoinitiators should have a quantum yield ϕ_{ph} as close as possible to one, which means that each photon initiates one reaction. However, a high ϕ_{ph} is not a guarantee for an efficient

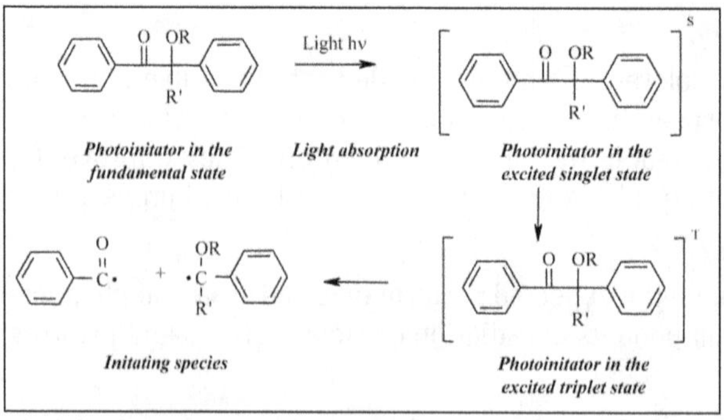

Figure 22: Type I photoinitiation (Norrish I reaction)

photoinitiator, since this value may also include competitive photoreactions which contribute to the conversion of the photoinitiator but do not result in the formation of initiating species. More significant for a radical photoinitiator is ϕ_r. In the case of Type I photoinitiator (Figure 22), two initiating radicals are formed and thus the maximum value of ϕ_r is two. In contrast, only one initiating radical is usually formed from a bimolecular Type II photoinitiator with a corresponding maximum value of ϕ_r of one (Figure 23). The closer to these limiting values the ϕ_r of an initiator is, the more efficient is the radical forming photoreaction.

Side reactions such as oxygen or monomer quenching decrease the value of ϕ_{ph}, which in reality is thus always lower than 1 (usually in the range of 0.4 to 0.8) and ϕ_r is also correspondingly lower than the theoretical value. Moreover, the radicals formed can undergo competitive reactions to the initiating process. These include radical recombination or disproportionation as well as trapping by reactive species such as oxygen dissolved in the formulation. Thus the quantum yield of initiating radicals ϕ_i, defined as the number of initiating radicals per photon absorbed, is mostly slightly lower than ϕ_r. Although this number is experimentally not easy to determine, it is however best suited to characterize the photoefficiency of a photoinitiator.

An attractive photoinitiator has to meet the following requirements:
- undergo chemical reactions upon the absorption of light
- high quantum yield of the photoreaction generating species which are able to initiate a polymerization process
- short lifetime of excited states (typically a few nanoseconds) in order to minimize quenching reactions
- a useful absorption spectrum with an optimum overlap with the light source emission
- good compatibility to other raw materials, easily dissolved in the formulation, and easy to handle
- provide a stable formulation in the absence of light
- thermally stable
- both the photoinitiator and its photoproducts have low odor and toxicity
- no or low migration of photoinitiator and photoproducts after UV exposure (by-products formed during the photoinitiator photolysis)
- decomposition products should not affect the weathering stability or other properties of the cured material
- cost efficient technical synthesis

In a second step, the choice of a photoinitiator by the end-user will be governed by the following conditions:
- irradiation time and intensity which necessity a certain photoinitiator reactivity
- exposure atmosphere
- required surface cure
- film thickness
- pigmented formulation or clearcoat

- yellowing acceptance
- required durability of the film

The following chapter covers the three major categories of photoinitiators currently available, i.e.:

- radical photoinitiators undergoing a homolytic scission (Type I)
- radical photoinitiators generating radicals via hydrogen abstraction (Type II)
- cationic photoinitiators

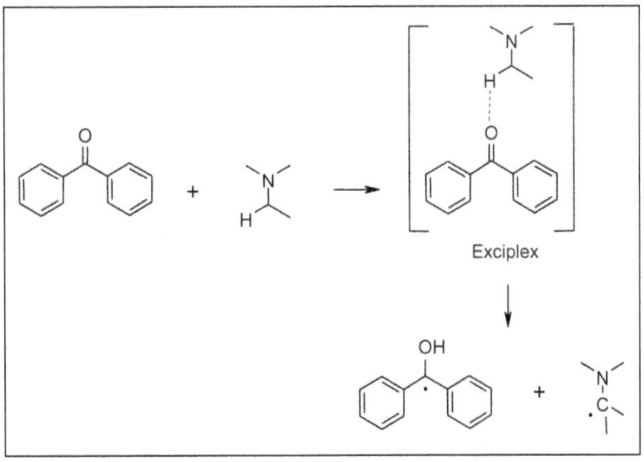

Figure 23: Type II photoinitiation by hydrogen abstraction

Detailed information on photoinitiators can be found in an excellent review from *Crivello* and *Dietliker* [4] and in a compendium of commercially available photoinitiators [5].

3.2.2 Free radical photoinitiators

Two families of free radical photoinitiators make up the largest part currently used in industrial applications [3]:

- Type I photoinitiators which undergo a unimolecular cleavage reaction (Norrish Type I cleavage) from the excited state as represented in Figure 22.
- Type II photoinitiators leading to the formation of radicals through hydrogen abstraction or electron transfer from a co-initiator, as shown in Figure 23. This initiation reaction is bimolecular as the presence of a hydrogen donor (usually an amine) is required in addition to the light sensitive photoinitiator.

3.2.2.1 Type I photoinitiators

Most of R&D efforts in developing new photoinitiators have been devoted to Type I photoinitiators since unimolecular processes are intrinsically more efficient than bimolecular reactions and many benzoyl-based chemical structures give access to very efficient α-cleavage photoinitiators. The following families of Type I photoinitiators have been developed to commercial compounds:

- benzoin derivatives
- benzile ketales
- α-hydroxyalkylphenones
- α-amino acetophenones
- acylphosphine oxides [6].

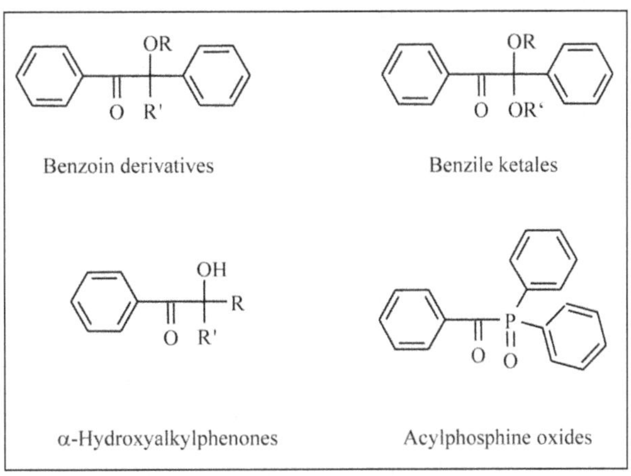

Figure 24: Basic structures of Type I photoinitiators

Basic structures of these families are represented in Figure 24: the chromophore of Type I photoinitiators is usually an aromatic carbonyl moiety which in the case of acylphosphine oxides further includes a phosphinoyl group.

Continuing developments have been governed by requirements from the coating and ink industry constantly demanding for improvements in terms of reactivity, coloration after curing, sensitivity to oxygen inhibition and efficiency in curing thick and/or pigmented films.

- **Benzoin derivatives**

Benzoin derivatives, in particular benzoin ethers are the first generation of Type I photoinitiators employed for curing vinyl monomers and silicone based formulations. In addition to poor initiating properties with acrylates, their low thermal stability due to the presence of benzylic hydrogen adversely affects the formulation potlife leading to premature gelling at room temperature. A further limitation of these components is a strong yellowing of the cured film. These multiple drawbacks associated with limited performance resulted in the substitution of these compounds by more performant photoinitiators. In fact, today benzoin ethers do not find any application in UV curing.

- **Benzile ketales**

The first Type I photoinitiator encountering a strong commercial success was 2,2-dimethoxy-2-phenylacetophenone (DMPA) [7], introduced in the market in 1977. Structure and light activation are represented in Figure 25, absorption spectra at different concentrations are shown in Figure 26. DMPA fulfils three of the most important photoinitiator requirements, i.e. a high thermal stability and a fast and efficient production of initiating radicals. Acrylate formulations containing DMPA are stable for years while an exposure of a few seconds to a medium pressure mercury lamp readily cures the film. Another feature of this compound is the very short lifetime of the excited triplet state. As a consequence, this compound is hardly quenchable and is the initiator of choice in styrene-containing formulations. In fact, other very efficient photoinitiators such as α-hydroxyalkylphenones (see below) are efficiently quenched by this monomer and are thus not suitable for these systems.

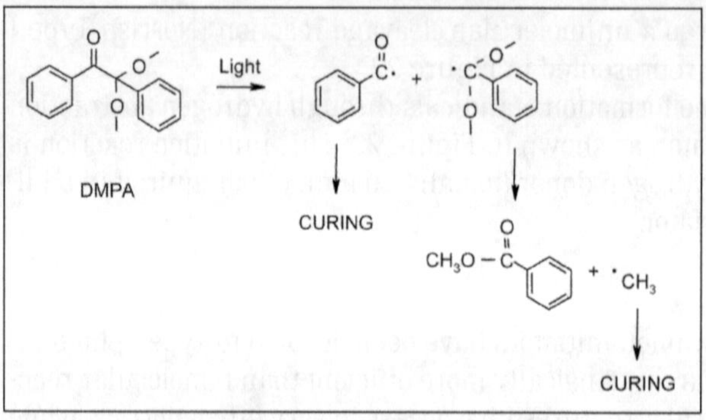

Figure 25: Photoinitiation by DMPA

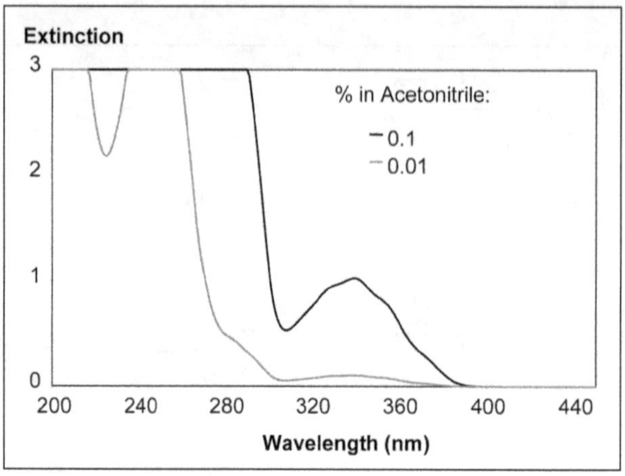

Figure 26: Absorption spectra at different concentrations of DMPA in acetonitrile - 1 cm thick cell

DMPA is still widely used since it is a very efficient low-cost photoinitiator. A major limitation of this photoinitiator is the strong film yellowing after curing [8, 9]: Figure 27 shows some undesired reactions occurring during irradiation of this photoinitiator. Some of these photoproducts such as benzil or quinoide structures are yellowish and thus supposed to be responsible for the observed yellowing.

Yellowing is tolerated in applications such as some wood coatings, printing inks and flexography inks but this issue largely justifies the development of new photoinitiators to cover needs where residual coloration is not acceptable, e.g. for clearcoats and outdoor coating applications.

More generally, a further drawback of photoproducts are their ability to migrate what is a major concern when considering food contact applications but also to a lower extent what also causes weathering issues.

- **α-hydroxyalkylphenones (or α-hydroxyacetophenones)**

In the search for non-yellowing photoinitiators, α-hydroxyalkylphenones were developed in the eighties [10]: the solid 1-hydroxy-cyclohexyl-phenyl-ketone (HCPK, Figure 28) and the liquid 2-hydroxy-2-methyl-1-phenyl-propan-1-one (Figure 29) were the first initiators allowing UV curing of clearcoats for commercial applications.

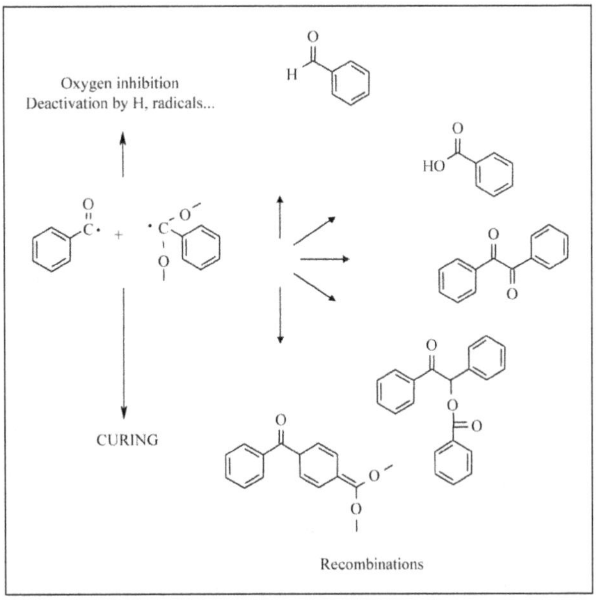

Figure 27: Possible reactions of radicals generated during DMPA light-activation

Yellowing is significantly reduced compared to DMPA [11] while they show a comparable reactivity. Additionally, they exhibit an improved solubility in acrylates than DMPA. The absorption spectrum of HCPK is represented in Figure 32.

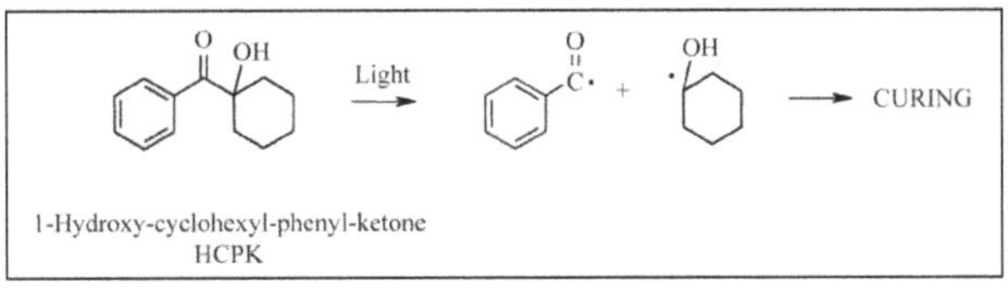

Figure 28: Photolysis of HCPK

Based on this structure, other derivatives were developed with focus on improved solubility, similar or enhanced reactivity and lower volatility. A compound that has been commercialized in 1992 is shown in Figure 29: the 2-hydroxethoxy substituent induces a red-shift in the absorption spectrum and considerably reduces the volatility. Moreover, the primary hydroxyl group allows for further chemical modification and imparts a modest solubility in polar media such as water. This compound has thus found use in water-borne formulations.

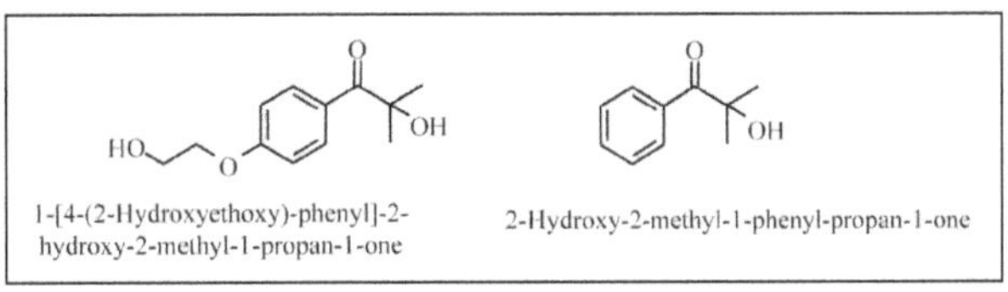

Figure 29: Examples of Type I α-hydroxyalkylphenones

To illustrate the high reactivity of HCPK in an acrylate formulation, the conversion of acrylate functions was monitored as a function of the exposure time at room temperature by real-time infrared spectroscopy [12]: the corresponding kinetic curve is represented in Figure 30 [13]. Both under air and inert atmosphere, initiation occurs within a fraction of a second while nearly complete conversion of acrylates is achieved within less than 5 seconds. As further explained in Chapter 3.3.1.3, the glass temperature of the final polymer film is too high to allow a complete conversion of the acrylates when UV curing is performed at room temperature, the sample vitrification reducing the chain mobility. However, low amounts of remaining acrylates after curing have little effect on the film properties.

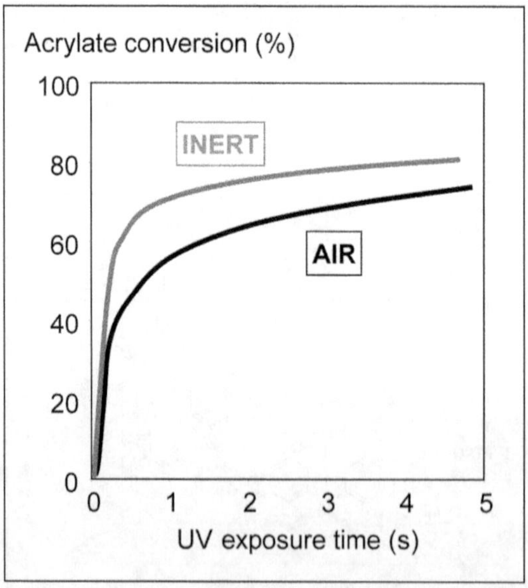

Figure 30: Acrylate conversion as a function of the irradiation time under air or inert atmosphere 3 wt% HCPK in a polyurethane acrylate, film thickness = 10 μm, exposed to a medium pressure mercury lamp - light intensity = 100 mW·cm^{-2} [13]

Absorption spectra represented in Figure 32 show that α-hydroxyalkylphenones can be activated by lamps emitting up to 380 nm. Photoproducts have an absorption similar to that of the photoinitiator which means that the overall coating absorption before and after light exposure remains similar. The absorption of the incident light by the photoinitiator or its photoproducts results in a reduction of the light intensity in deeper layers of the coating. Since the light intensity decreases exponentially with the penetration depth (Lambert-Beer's law), the light is absorbed mostly in the surface layer at wavelengths where the photoinitiator has a strong absorption, thereby preventing this radiation from reaching the bottom of the coating. This phenomenon is called **inner filter effect**, which is most pronounced in the UV-C and UV-B region and may impeach an efficient through-cure. α-Hydroxyalkylphenones are therefore recommended for UV curing of thin clearcoats or for surface curing, and confer satisfying surface properties to the final material.

- **α-amino acetophenones**

Pigmented formulations such as offset or flexography inks and photoresist applications are cured by light at longer wavelength (e.g., monochromatic light in case of i-line) and require different photoinitiator absorption characteristics than clearcoats. α-Amino acetophenone photoinitiators were developed especially for these markets. The chemical structures of the two most important α-amino acetophenones are represented in Figure 31 [14]. Based on similar acetophenone-type structures as α-hydroxyalkylphenones, the exchange of the hydroxy group by an amino substituent confers specific properties to these photoinitiators, and the introduction of an electron donating substituent on the benzoyl moiety strongly modifies the absorption spectrum of the molecule and shifts it into the visible region as shown in Figure 32. Due to this broad absorption in the UV-A region, such photoinitiators exhibit an improved reactivity compared to α-hydroxyalkylphenones in the

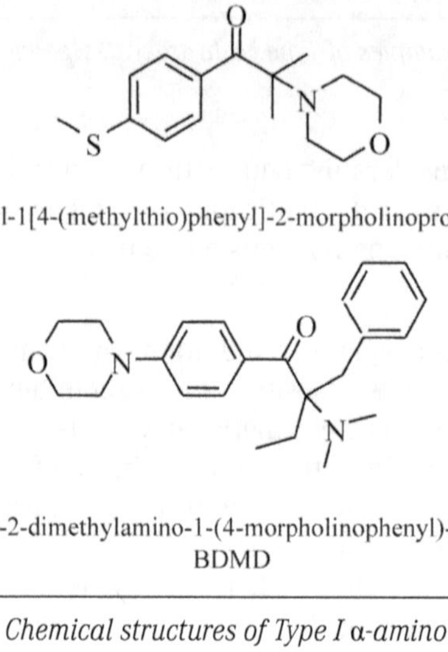

2-Methyl-1[4-(methylthio)phenyl]-2-morpholinopropan-1-one

2-Benzyl-2-dimethylamino-1-(4-morpholinophenyl)-butanone-1
BDMD

Figure 31: Chemical structures of Type I α-amino acetophenone

presence of pigments or when activated with light of longer wavelengths and are primarily recommended for ink and photo-resist applications.

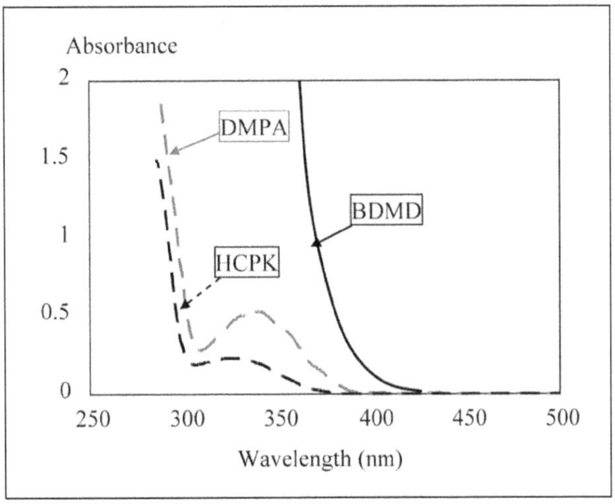

Figure 32: Comparison of the absorption spectra of DMPA, HCPK and BDMD - 0.05% in acetonitrile

An additional benefit resulting from the presence of tertiary amino groups is a lower sensitivity to oxygen inhibition [15] thus improving reactivity and cure efficiency under air compared to HCPK: indeed, it has been proven that the addition of tertiary amines to UV curable formulations significantly reduces oxygen inhibition by converting non-reactive peroxy radicals into hydroperoxides and active α-aminoalkyl radicals [13], as shown in Figure 33.

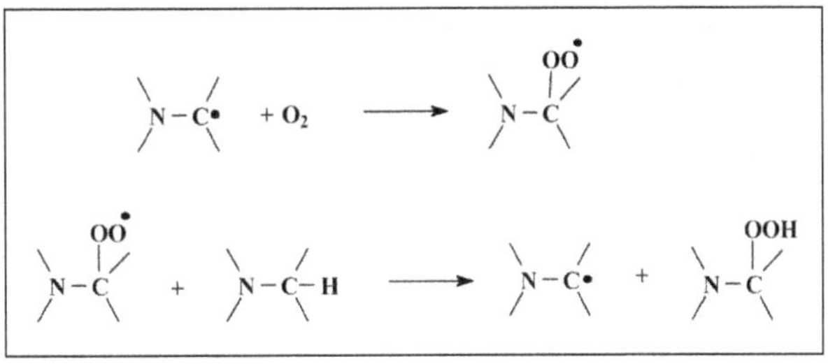

Figure 33: Trapping of oxygen by tertiary amines during a UV curing process

Since printing inks are used in very thin layers, curing of these systems is likely to be strongly affected by oxygen inhibition. α-Amino acetophenones are therefore recommended for these applications but are not adapted for clearcoats and outdoor long-term utilization: inherently to oxygen scavenging, amine oxidation also adversely affects the ageing properties of the cured material by imparting a strong yellowing.

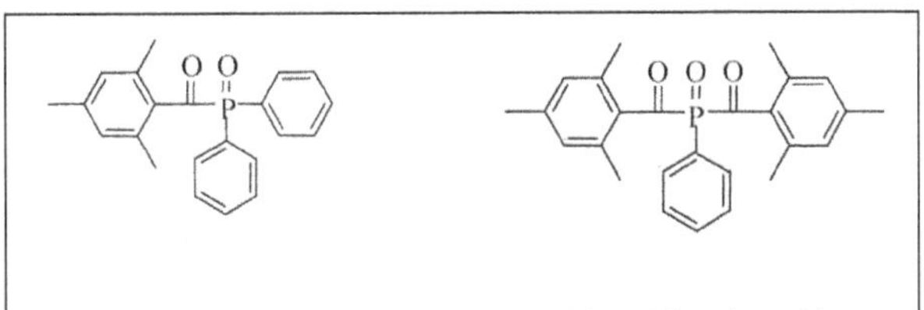

Monoacylphospine oxide Bisacylphospine oxide

Figure 34: Examples of mono- and bis-acylphosphine oxides

- **Acylphosphine oxides**

The most recent category of Type I photoinitiators introduced in the market are mono- and bis-acylphosphine oxides represented in Figure 34 and respectively called MAPO and BAPO. These components were a major breakthrough in UV curing technology as their structure is fundamentally different from previous photoinitiators and brought some unique features, among the most characteristic ones:

- high reactivity providing a high cure speed
- strong absorption in the UV-A extending well into the visible, allowing excitation in the presence of components absorbing and scattering strongly in the UV-A region, such as white pigments or UV absorbers
- photobleaching properties allowing the continuous penetration of light in deeper layers in thick and pigmented films, as well as providing non-yellowish cured coatings

The reactivity of acylphosphine oxides can be readily explained by the outstanding initiating ability of phosphinoyl radicals versus acrylates [16, 17]. An additional feature of BAPOs

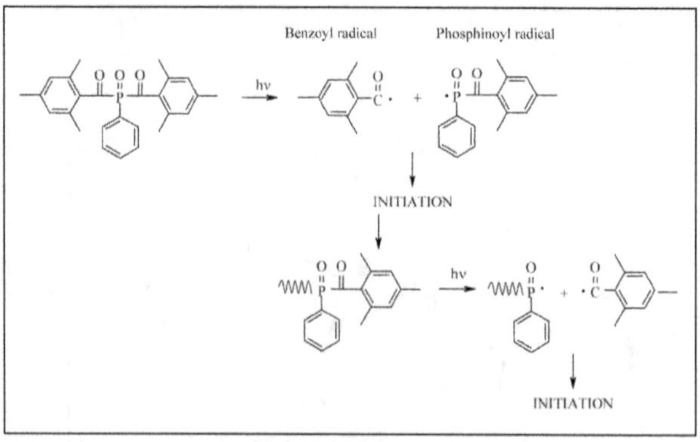

Figure 35: Photolysis of BAPO by two photons

is the generation of four radicals per photoinitiator molecule, which means that one molecule of BAPO can initiate four chains by absorbing two photons as shown in Figure 35.

The broad and strong absorption of acylphosphine oxides in the near-visible/visible region (380 to 420 nm) is shown in Figure 36: metal doped medium pressure mercury lamps and fluorescent light sources can readily activate these photoinitiators. The chromophore based on a phosphinoyl group conjugated to a benzoyl moiety confers a yellow coloration to the photoinitiator. Since this group is destroyed during photolysis (Figure 35), the optical density of the coating in the UV-A region significantly differs before and after light exposure, and during irradiation progressively shifts to shorter wavelengths: films bleach during curing (figure 37). This phenomenon is called **photobleaching** and continuously reduces the inner filter effect of the film, thus allowing the deep and efficient penetration of UV-A radiation through thick clearcoats as illustrated in Figure 38: films up to several centimeters can be cured using acylphosphines oxides which are also well adapted for pigmented coatings. Figure 39 shows the filter effect of TiO_2 and confirms that there is a window between 400 and 420 nm available for activating BAPO in white pigmented formulations: 100 µm thick white films containing BAPO can be cured by UV light.

Acylphosphine oxides provide efficient though-cure and are therefore recommended for the fast curing of thick and/or pigmented sections [18]. They are well adapted for the curing of acrylates, styrenes, unsaturated polyesters and thiol-polyene resins. Remaining non-photolyzed photoinitiator may confer a slightly yellow coloration to the final material: an extended light exposure usually remedies to this issue. Visible light, e.g. sunlight, can activate acyl-

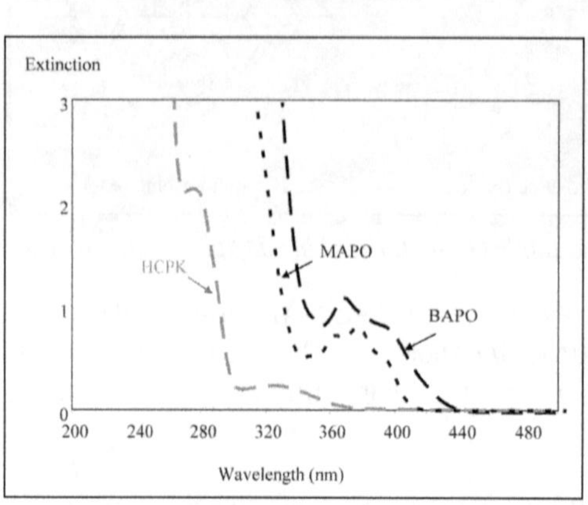

Figure 36: Comparison of the absorption spectra of a BAPO (0.05% in acetonitrile), MAPO (0.05% in methanol) and HCPK (0.05% in acetonitrile), 1 cm thick cell

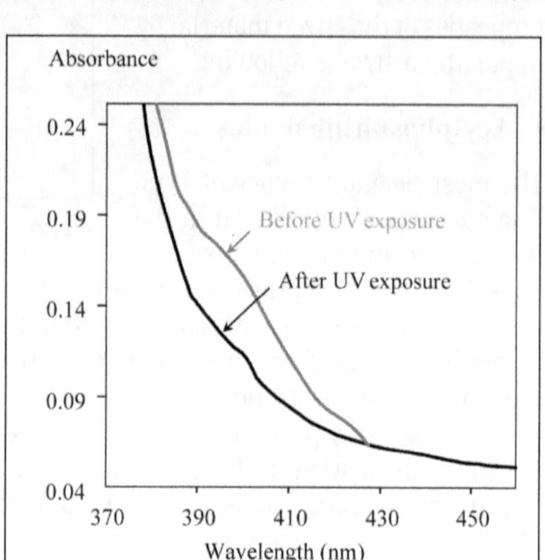

Figure 37: UV absorption of BAPO in a polyurethane acrylate before and after UV exposure 1 wt% BAPO, 12 µm thick film, UV dose = 500 mJ/cm²

phosphine oxide photoinitiators: it is mandatory to protect the formulation from ambient light and to store it in opaque bottles.

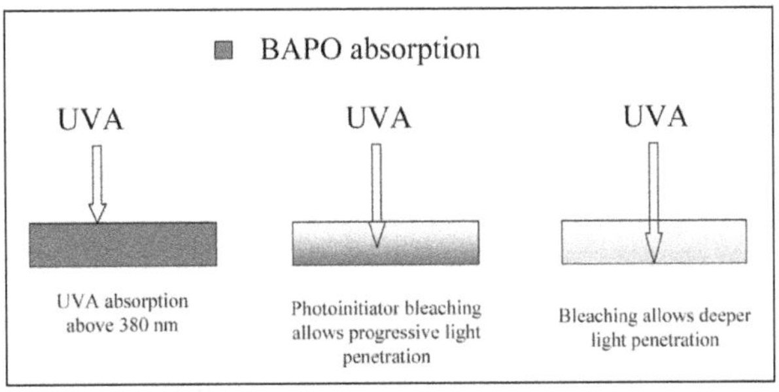

Figure 38: Photobleaching of acylphosphine oxides allowing a deep through-cure

- **Phenylglyoxylates**

Other monomolecular photoinitiating processes used for the formation of initiating radicals are based on the Norrish Type II reaction which leads to the formation of 1,4-biradicals under the action of UV light via an internal hydrogen abstraction from the γ position to the carbonyl group as represented in Figure 40. Phenylglyoxylates undergo such Norrish Type II photochemical process. In a radiation curable formulation, this biradical can undergo bimolecular reactions with the components of the formulation besides the usual cleavage to give a hydroxy ketene and an aldehyde which both are not initiating species. Initiating radicals are suggested to be formed by multistep as shown in Figure 40 [19, 20].

An attractive property of phenylglyoxylates is their low yellowing in clearcoats which may be attributed to the fact that no radical cleavage products are formed, which could recombine to give colored photoproducts. Additional advantages are an excellent surface cure and no yellowing upon ageing. Due to their reduced reactivity compared to α-hydroxyalkylphenones, these components have encountered a lower commercial success.

3.2.2.2 *Type II photoinitiators*

Type II photoinitiators are essentially based on aromatic ketones (e.g., benzophenone, thioxanthones or camphorquinones [21]) and undergo after light absorption a bimolecular reaction mostly involving electron transfer followed by proton abstraction from a co-initiator, usually a tertiary amine [22]. Figure 41 illustrates an example of Type II photoinitiation with benzophenone: polymerization initiation occurs through the amino alkyl radical while the ketyl radical is not active towards double bonds.

Moreover, amines help to decrease oxygen inhibition but impart a strong yellowing to the film, poor weathering stability and have a softening effect on the final material [3], as

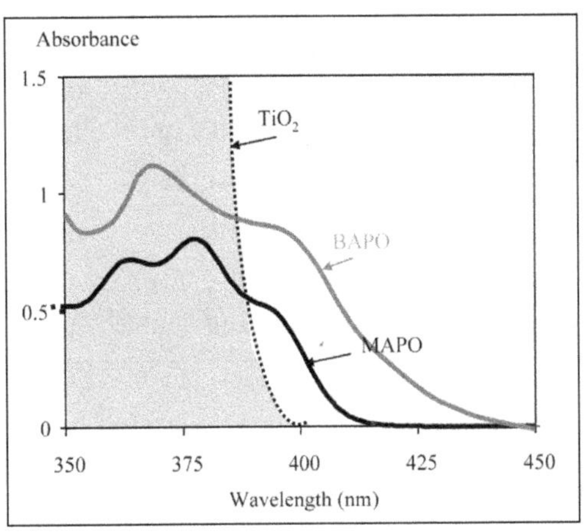

Figure 39: TiO_2, MAPO and BAPO absorption in a polyurethane acrylate

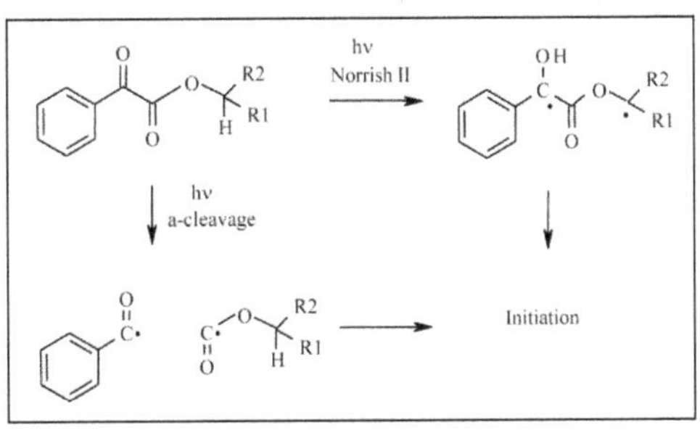

Figure 40: Photoinitiation mechanism of phenylglyoxylates through combination of Norrish I and Norrish II reactions

Table 6: Characteristics of the different classes of Type I photoinitiators

	Benzoin derivatives	**Benzile ketales**	**α-Hydroxyalkyl phenones**	**α-Amino acetophenones**	**BAPO & MAPO**	**Phenyl glyoxylates**
reactivity	low	high	high	extremely high	very high	moderate
oxygen inhibition	strong	moderate	moderate	very low	low	moderate
bleaching	no	no	no	no	yes	no
yellowing	strong	strong	low	strong	medium to low	low
cure	surface cure	surface cure	surface cure	surface cure and through cure of pigmented systems	through-cure and pigmented systems	surface cure
typical concentration*	/	0.5 to 4 %	0.5 to 6 %	0.5 to 4 %	0.1 to 4 %	1 to 4 %
application	none	wood inks	clearcoats paper coating	offset, flexography, inks, photoresists	thick pigmented films & clearcoats	clearcoats
coating ageing resistance – coloration	poor	poor	excellent	poor	medium	outstanding
commercially available	no	yes	yes	yes	yes	yes

** depending on the application*

explained p. 37. In spite of these major drawbacks, this low cost system still occupies an important position in the UV curing market for low quality articles.

Note

Oxygen inhibition can also be reduced to a certain extent by the use of ethylene or propylene glycol. It is proposed that the mechanism of -O-CH is similar to the -N-CH-group.

Table 6 summarizes distinct characteristics of the different classes of Type I photoinitiators.

3.2.2.3 Photosensitizers

A photosensitizer useful for UV curing enhances the spectral response of a formulation by extending the sensitivity of the photoinitiator to longer wavelengths. In other terms, it absorbs part of the light the photoinitiator cannot absorb and transfers the energy to the photoinitiator. Afterwards the sensitizer comes back to its ground state and can re-activate other photoinitiator molecule:

$$\text{Sensitizer} \xrightarrow{h\nu} \text{Sensitizer}^* \xrightarrow{PI} PI^* + \text{Sensitizer.}$$

Some well-known sensitizers are based on a thioxantone structure, e.g. ITX. A more explicit illustration is given in Figure 42 which shows the absorption spectra of sensitizer ITX and an α-amino acetophenone: the sensitizer absorbs radiation comprised between 350 and 400 nm and transfers the corresponding energy to the photoinitiator. Sensitization is useful in the case the photoinitiator absorp-

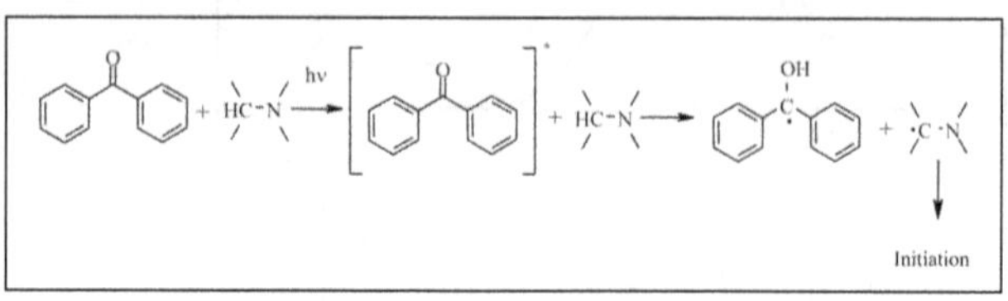

Figure 41: Photoinitiation mechanism by hydrogen abstraction

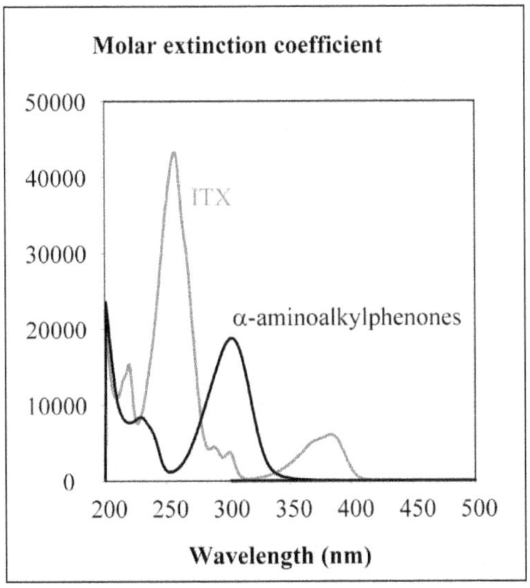

Figure 42: Absorption spectrum of ITX and α-aminoacetophenone

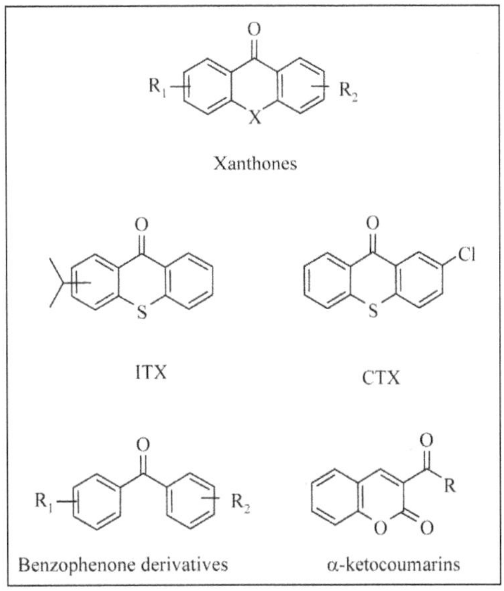

Figure 43: Examples of sensitizers (X = S or O)

tion is limited to deep-UV (below 300 nm) or for pigmented formulations, but also for UV-A curing, UV-C and UV-B radiation being a major concern for health hazard.

This process obeys fundamental physical laws which stipulate that energy transfer is only possible from a state of higher to one of lower energy. For an efficient process, sensitizer excited state has to be higher than that of the photoinitiator by at least 3 kcal.mol^{-1} [4]. Therefore an appropriate selection of a sensitizer/initiator pair which meets this energetic requirement is necessary in order to design an efficient initiating system. Since the triplet state of most α-amino acetophenones is of relatively low energy, these Type I photoinitiators can be readily activated by sensitizers absorbing at long wavelengths. Thioxanthones and more generally xanthones, benzophenone derivatives, anthracenes and ketocoumarins are some categories of sensitizers shown in Figure 43. Concentration usually ranges between 0.2 and 3 % with ITX, 2 and 8 % with benzophenone.

3.2.2.4 Visible UV curing

Efficiency of dye/coinitiator systems (e.g., rose bengal) and photoinitiators (e.g., titanocene) absorbing in the visible region has also been investigated for curing by visible light. Photoinitiators absorbing in the visible are colored compounds and require handling under yellow or red light in order to prevent premature gelation. These properties make visible photoinitiators less suitable for coating applications and their use is restricted to other high value niche applications. They find use in high-tech applications such as laser direct imaging using visible lasers or for electronic materials. Camphorquinone is another visible photoinitiator that is mainly used for the curing of dental materials, since UV light can not be used in this application due to the possible damage of the mucosa.

3.2.3 Cationic photoinitiators

Although of lower commercial importance than free radical photoinitiators, cationic photoinitiators have attracted a lot of interest (*Crivello* et al. [23, 24]). The following list summarizes the main categories of cationic photoinitiators producing strong *Brönsted* or *Lewis* acids upon

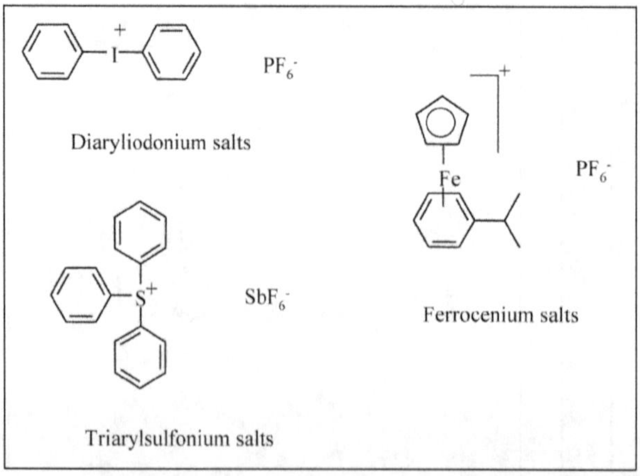

Figure 44: Chemical structures of some cationic photoinitiators

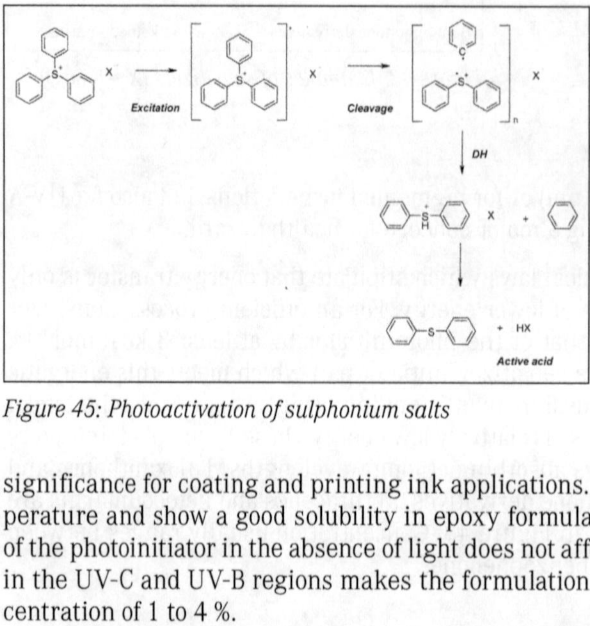

Figure 45: Photoactivation of sulphonium salts

irradiation, which can be used for curing epoxies and vinyl ethers by UV light (structures shown in Figure 44):

- triarylsulphonium salts
- diaryliodonium salts
- ferrocenium salts
- diazonium salts
- dialkylphenylacylsulphonium salts

α-Sulfonyloxy ketones and oxime sulfonates are other efficient photoinitiators that generate sulfonic acid derivatives upon irradiation. While these weaker acids are not suitable for the curing of epoxides or vinyl ether derivatives, they find use as photoacid generators (PAG) for crosslinking or deprotection reactions mainly used in microelectronic applications. The very specific requirements of these applications resulted in the development of a series of tailor-made PAGs that are commercially available.

Onium salts (triarylsulphonium and diaryliodonium salts) are the only ones to have a commercial significance for coating and printing ink applications. They are usually stable at room temperature and show a good solubility in epoxy formulations, while the intrinsic low acidity of the photoinitiator in the absence of light does not affect the resin potlife. Their absorption in the UV-C and UV-B regions makes the formulation stable under daylight at typical concentration of 1 to 4 %.

A simplified illustration of sulphonium salt photolysis is shown in Figure 45: the photogeneration of acidic species HX initiates the cationic curing of epoxy or vinyl oligomers. Weak *Brönsted* acids are not suitable for cationic curing while strong *Lewis* and *Brönsted* acids are well adapted and confer a high reactivity to the photoinitiator. Generation of benzene has been a crucial point that can be overcome by the introduction of substituents on the phenyl ring.

In the case of iodonium salts, acid generation proceeds as following:

$Ar_2I^+X^- \rightarrow [Ar_2I^+ X^-]^*$

$[Ar_2I^+X^-]^* \rightarrow ArI^{\bullet +} + X^- + Ar^{\bullet}$

$ArI^{\bullet +} + X^- + DH \rightarrow ArIH^+ + X^- + D^{\bullet}$

$ArIH^+ + X^- \rightarrow ArI + HX$

Figure 46 shows some iodonium and sulfonium salts which are currently commercialized.

The counter ion X plays a key role in the initiation process in that it controls the strength of the acid generated by the photolysis, thus determining the cure speed. Typical anions for cationic photoinitiators in the order of increasing efficiency are BF_4^-, PF_6^-, AsF_6^- and SbF_6^-.

Numerous attempts have been made to modify the absorption spectra of sulfonium salts and enhance their low absorption in the 300 to 400 nm region essentially by introducing extended aromatic chromophores. A limited choice of cationic salts absorbing above 340 nm is currently commercially available, while photosensitizers are often added to formulations containing simple onium salts in order to shift the absorption towards longer wavelengths (Figure 47). Anthracenes readily sensitize sulphonium salts, while a more ample variety of sensitizers is available for iodonium salts, which includes typical photosensitizers such as thioxanthones, benzophenones, anthracenes, perylenes and xanthones. Some photoinitiators, e.g. α-hydroxyalkylphenone HCPK also manage to activate iodonium salts albeit by a different mechanism, while keeping the yellowing at a relatively low level after UV exposure.

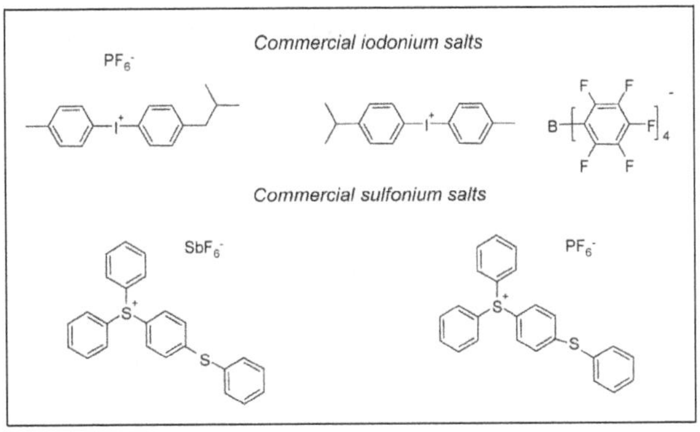

Figure 46: Commercialized cationic photoinitiators

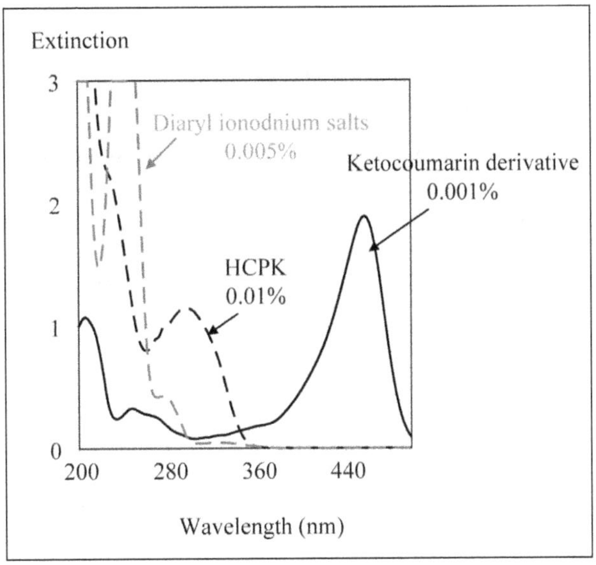

Figure 47: Absorption spectra of a diaryliodonium salt, a ketocoumarin derivative and HCPK in acetonitrile

Some recent developments address solubility and photoproduct toxicity issues by incorporating alkyl chains to the phenyl groups, while PF_6^- counterion provides a satisfying compromise in terms of reactivity and heavy metal-free photoinitiator. These products have found many applications in UV curing of cycloaliphatic epoxies and vinyl ethers for white coatings, UV flexo inks, UV screen inks, overprint varnishes and adhesives.

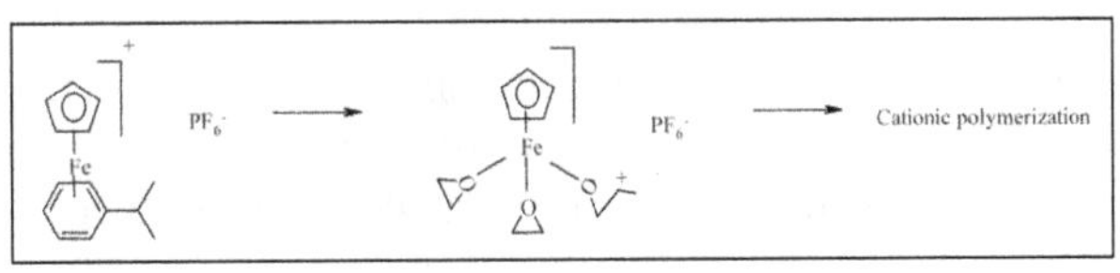

Figure 48: Photoactivation of ferrocenium salts

Activation of ferrocenium salts to generate initiating *Lewis* acid is illustrated in Figure 48. Such cationic photoinitiators are still commercialized in small amounts for cationic curing by visible light.

In some niche applications such as photoimaging and photoresists, cationic photoinitiators have found applications in initiating a depolymerization process followed by evaporation of the resultant monomer [25, 26]. Aryldiazonium salts of complex halides releasing *Lewis* acids under light did not encounter any commercial success, but have attracted some attention [27].

3.2.4 *Photoinitiators for clearcoats*

The selection of one photoinitiator rather than another one as well as the determination of the required concentration are governed first by the specifications the final material has to fulfill (yellowing, thickness, durability) as well as by the curing conditions (intensity, lamp emission, curing atmosphere). A key element one has to take into consideration as well is the resin absorption (inner filter effect of clearcoat or absorption of pigmented films) which limits the depth of penetration of light throughout the film.

Regarding clearcoats, following film properties will be significantly influenced by the choice of photoinitiator and the curing conditions. However, it has to be kept in mind that most of these properties are also strongly influenced by other components of the formulation, such as the resin type or additives that are incorporated:

- scratch resistance → surface cure
- tackiness → surface cure
- adhesion to the substrate → through-cure
- hardness, chemical & mechanical resistances, weathering stability → overall cure
- yellowing → photoinitiator type and concentration

Depending on the applications, the following photoinitiators can be recommended for clearcoats:

- **α-hydroxyalkylphenones** confer excellent surface cure, good reactivity, limited oxygen inhibition, little yellowing and good weathering stability to the final material
- ***phenylglyoxylates*** provide good surface properties with no yellowing, no odor and good weathering stability
- highly reactive **acylphosphine oxides** have to be used for thick clearcoats (>20 µm) to enhance the through-cure (typically 0.1 to 1 %). They are efficient at a low level with little yellowing
- **blends of acylphosphine oxides with α-hydroxyalkylphenones** are commercially available for high end value applications, e.g. in the automotive industry, and have been optimized in terms of balance between surface/through-cure and yellowing. α-hydroxyalkylphenones provide the surface cure (UVB-light absorbed at the surface) while photobleaching allows the activation of acylphosphine oxides in deep-lying layers by UVA-light
- **benzophenone** combined to amines can be used for low cost and poor quality applications where yellowing during ageing is not an issue. However, combination of benzophenone and α-hydroxyalkylphenones has shown to be quite reactive and little sensitive to oxygen inhibition. Optimized blends are commercially available
- **benzile ketales** impart a considerable yellow coloration but give high cure speed at a low cost. They are also useful in styrene-containing formulations
- most of **sensitizers** (thioxanthones, anthracenes) are usually not recommended in clearcoats as they confer a strong coloration to the film and do not exhibit major benefits compared to acylphosphine oxides. Excepted price, their use is therefore quite marginal in free-radical curing of clearcoats, excepted benzophenone. Cationic curing is also an additional exception as photosensitizers enhance the spectral response of the film.

3.2.5 *Photoinitiators and photosensitizers for pigmented coatings*

Main limitation of UV curing with pigmented systems is the competitive absorption of incident light by both the pigment and the photoinitiator. Pigment light absorption and scattering decrease the amount of light available for curing deep-lying layers thus directly affecting the film properties: the photoinitiator absorption has to overlap with the transmission window of the pigment and the emission spectrum of the light source to reach an efficient through-cure.

A careful selection of the photoinitiator package has to be realized, as sufficient light has to be available both in the surface and the bottom layers in order to allow efficient curing. The following photoinitiating systems are therefore recommended for pigmented systems:

- mixtures of **α-hydroxyalkylphenones** and **acylphosphine oxides** allow thickness up to 100 μm of white pigmented films to be cured with good surface properties and through-cure (typical ratio α-hydroxyalkylphenones and acylphosphine oxides is 1:1)
- **α-aminoacetophenones** confer a high reactivity and low oxygen sensitivity to pigmented systems, and are recommended for black, cyan, magenta and yellow formulations. Addition of a sensitizer (e.g. ITX) improves even more the formulation reactivity
- **propoxy-substituted thioxanthones** mixed to **tertiary amines** are used in offset applications, but such systems show a reduced cure speed.

UV curing of pigmented systems has found its main applications in inks (offset, screen, flexo or inkjet inks) with films thinner than 10 μm rendering UV curing with the above mentioned photoinitiating systems easy. Surprisingly, it has been observed that UV curing of pigmented films is less affected by oxygen inhibition than clearcoats, in particular at the film surface: this effect comes from light scattering by pigment particles which increases the amount of light available for surface curing [28].

3.2.6 New developments

The UV curing industry is constantly demanding for innovative solutions to increase productivity and improve film properties while reducing costs. Customers recurrently ask for new photoinitiators which could overcome some well-known issues listed as following:

- oxygen inhibition ↔ increased photoinitiator reactivity
- odor after UV exposure
- evaporation of the photoinitiator during flash-off step prior to UV irradiation
- migration of non-reacted photoinitiator, causing yellowing and premature ageing
- migration of photoproducts and non-photolyzed photoinitiator causing health hazard
- UV curing of thick pigmented films
- yellowing immediately after UV exposure
- yellowing during weathering

  - oligomeric photoinitiators have consequently been developed as low emission photoinitiators in the early 90s: some lead candidates have been quite successful but reactivity is still not completely satisfying. "Polymeric ITX" and benzophenone currently commercialized show mitigated benefits in terms of cure speed and are still an open point looking for a solution
  - with the same objective, strong progresses are currently being realized for high molecular weight photoinitiators for radical curing
  - dimeric α-hydroxyalkylphenone has recently been introduced to the market to provide a low emission and low odor photoinitiator. Contrary to oligomeric photoinitiators, reactivity is improved compared to the corresponding α-hydroxyalkylphenone [29]
  - dimeric phenylglyoxylate has also been launched as low odor and low yellowing photoinitiator: it exhibits a similar reactivity compared to its monomer counterpart [30]

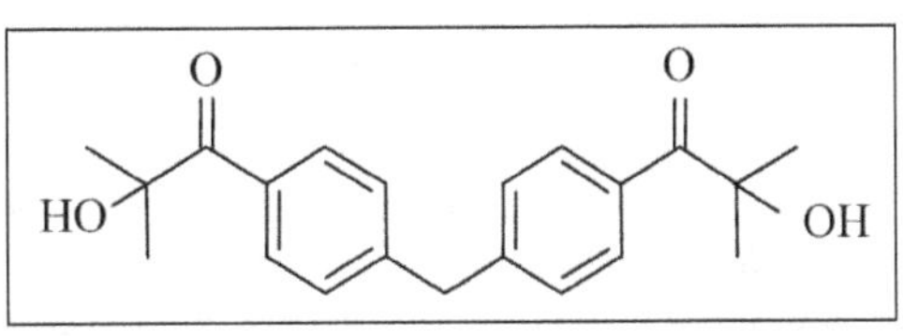

Figure 49: Dimeric α-hydroxyalkylphenone

- water-borne UV curable formulations are gaining importance since a few years: the photoinitiator industry correspondingly developed dispersions of acylphosphine oxide well incorporable to binder dispersions [31]
- for improving surface cure, photoinitiator migrating to the surface have been developed to increase the local concentration at the film surface [32]
- major limitation of cationic photoinitiators is their low absorption in the UV-A region requiring the addition of sensitizers: some efforts are currently being devoted to develop red-shifted cationic photoinitiators [33]

3.3 Monomers and oligomers

Monomers and oligomers play a major role in determining the physical properties of a radiation curable (RC) formulation and the resulting film. Regardless of their chemical structure, they always require at least one polymerizable group. In the case of curing by a free-radical polymerization mechanism, which can be initiated by electron beams or via radical-generating photoinitiators, the polymerizable groups are in general carbon-carbon double bonds. Monomers and oligomers with acrylate functions have become standard, mainly because of their high reactivity. Cationically curable substances often contain epoxy groups, and less frequently vinyl ether and oxetane groups [34, 35, 36] (see Chapter 2.1.2).

The next section describes the properties of reactive diluents and resins, and their effects on particular properties of RC formulations. Because the transition from monomers to oligomers is fluid, no distinction will initially be made between them. Systems cured by the mechanism of free radical polymerization are of significantly greater importance compared to cationic formulations, and a number of the examples presented below refer to these. In general, the relationships described here apply also to cationic polymerization.

Formulators need to modify the physical characteristics of liquid formulations and cured films. For this reason, generally applicable characteristics are collected in Table 11 to facilitate in the selection of appropriate raw materials from the variety of products available. Finally, the most important classes of monomers and oligomers and their specific properties are discussed.

3.3.1 General structure-properties relationships

3.3.1.1 Functional groups and functionality

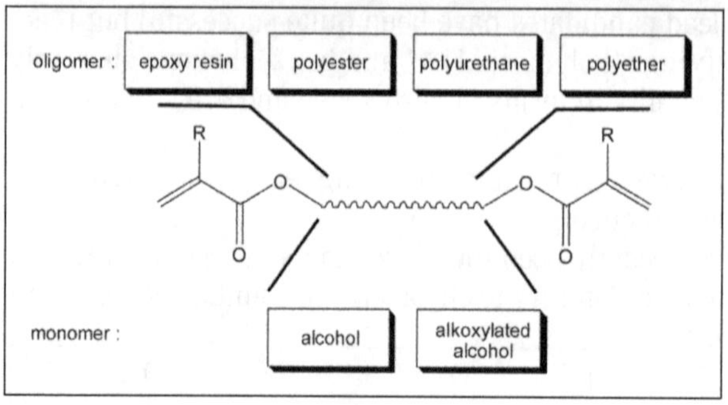

Figure 50: Schematic chemical structure of radically curable acrylate monomers and oligomers with two polymerizable groups; R = H (acrylate), CH_3 (methacrylate). The chain may be, e.g., a (cyclo)aliphatic alcohol, an oligo- or polyether, an epoxy resin, a polyurethane, or a polyester

The most important radically curable monomers and oligomers used in RC coatings and inks contain acrylate or, less frequently, methacrylate groups [37]. With the exception of these polymerizable groups, which are generally located at the end of the oligomer backbone, the chemical structures of reactive diluents and binders can vary widely. Oligomers, for example, can consist of polyesters, polyethers, or epoxy resins of relatively low molecular weight. The monomers are usually derived from monoalcohols, diols, or polyols that are sometimes alkoxylated. Figure 50 shows the basic structure of such a molecule. The term (meth)acrylate will be used below to refer to the acrylate as well as the methacrylate functional groups. The double bonds in unsaturated polyesters, on the other hand, lie along the backbone of the polymer (Figure 51).

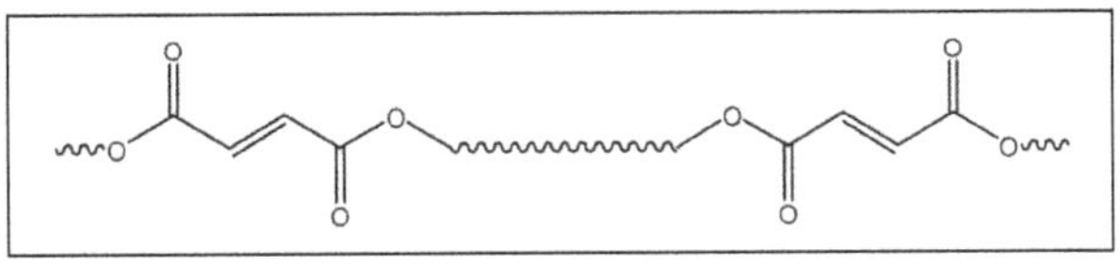

Figure 51: Schematic chemical structure of a unsaturated polyester containing fumaric acid

Cationically curable monomers and oligomers generally carry epoxy groups derived from, for example, epoxidated (cycloaliphatic) olefins (Figure 52). Vinyl ethers or oxetanes can also be cured cationically and are usually used in combination with epoxies to modify the properties of the systems.

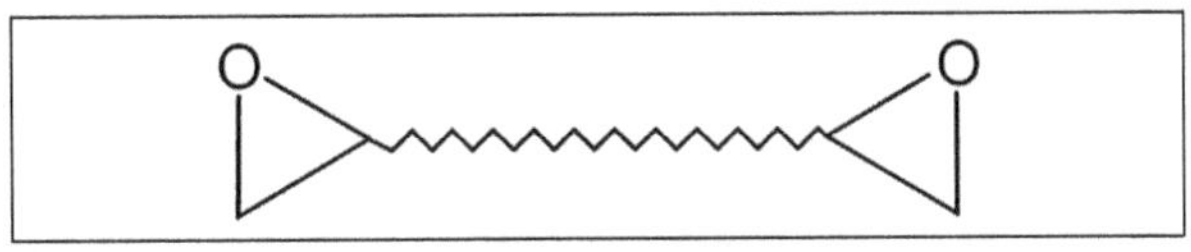

Figure 52: Schematic chemical structure of a cationically curable monomer or oligomer with two polymerizable epoxy groups; the chain may be, e.g., (cyclo)aliphatic

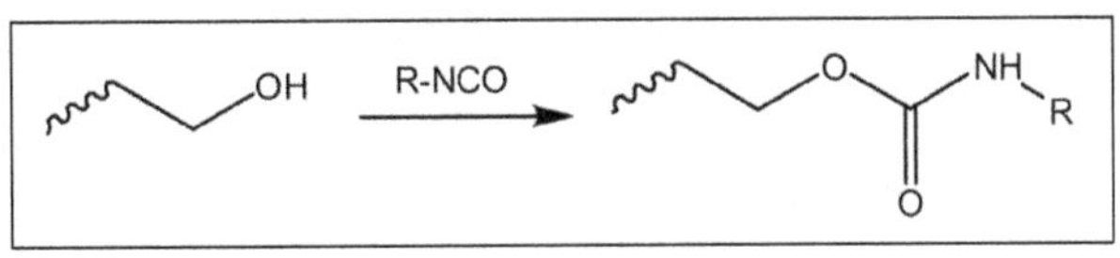

Figure 53: In conventionally curable systems, monofunctional raw materials are chain terminators

Apart from the type of polymerizable group, the functionality, i.e., the number of polymerizable groups per molecule, is of major importance. Conventional, thermosetting coatings crosslink via step polymerization. They therefore require higher-functional reactants to obtain durable and crosslinked films. If monofunctional compounds such as monoalcohols and monoisocyanates are used here, chain termination occurs, resulting in poor resistance properties (Figure 53).

In contrast, radical or cationic polymerization occurs via chain propagation: monofunctional and difunctional molecules give rise respectively to linear and tetrabranched polymer structures. The crosslink density is therefore much higher than in conventional systems, resulting in films with extraordinarily high hardness and chemical and scratch resistance (see Chapter 3.3.1.4.2). Figure 54 illustrates this, using the example of polymerizing acrylates. As seen in the figure,

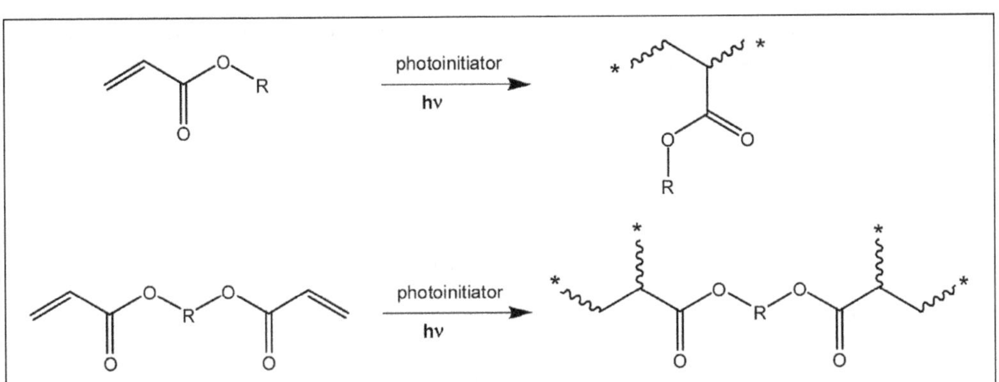

Figure 54: Functionality and crosslinking of radically curable monomers and oligomers. Monofunctional compounds form linear chains; difunctional substances form tetrabranched polymer structures; R = e.g., (cyclo)aliphatic group,
** = growing polymer chain*

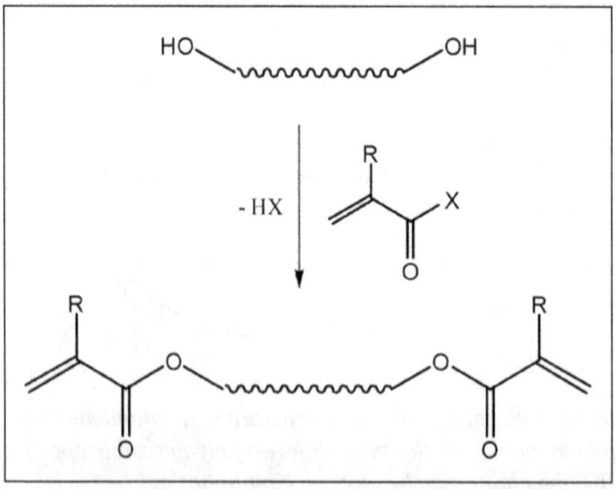

Figure 55: Preparation of a (meth)acrylate-containing monomer/oligomer with two polymerizable groups; R = H, CH_3; chain: (cyclo)aliphatic, oligo- or polyether, epoxy resin, polyurethane, or polyester, etc; X = OH, OCH_3, $OCOCH=CH_2$

the crosslink density depends upon the average number of polymerizable groups per molecule (i.e., the functionality) and the molecular weight between two crosslinks. The functionality normally lies between one and six for monomers, and between two and six for oligomers.

Many different scenarios exist for preparing radically curable (meth)acrylic monomers and oligomers, but all follow the general principle of inserting a polymerizable group after synthesis of the starting compound. Starting materials with hydroxy or epoxy functional groups are, e.g., reacted with (meth)acrylic acid in such a way that the double bonds remain intact. Figure 55 illustrates this method using the example of esterification of a diol with acrylic acid (or a derivative thereof).

3.3.1.2 Influence of monomers and oligomers on the viscosity of the formulation

In RC coatings and printing inks, volatile solvents are generally not used at all. The viscosity of the individual components must therefore be as low as possible to ensure satisfactory production as well as good flow behavior of the liquid formulation. The viscosity of the monomers and oligomers depends on a variety of factors, summarized in Table 7.

The most important parameter is the molecular weight. With increasing molecular weight the hydrodynamic volume increases, leading to reduced mobility of the molecules and hence to higher viscosity. In general, the number average molecular weight (M_n) of monomers lies below 500 g/mol and that of oligomers between 500 and 5,000 g/mol, allowing attainment of low viscosities [37].

Note

The frequently advanced explanation that the number of entanglements increases with increasing molecular weight does not apply in the present substances because their molecular weights are too low.

Table 7: Influence of monomer and oligomer properties on viscosity

Properties leading to low viscosities	Explanation
low molecular weight	low hydrodynamic volume
few polar groups (-OH, -COOH, urethane, etc.)	less intermolecular hydrogen bonding
low polydispersity (Mw/Mn)	lower proportion of particularly high-molecular compounds
high degree of branching	low hydrodynamic volume for the same molecular weight
good dissolving power and good solubility/compatibility	formation of gel-like structures prevented
low glass transition temperature	low hydrodynamic volume

Oligomers (as well as many monomers, see Chapter 3.3.3) have a wide molecular weight distribution rather than an exact molecular weight, because they are usually mixtures of dimers, trimers, tetramers, etc. The width of the distribution is described by the polydispersity, which expresses the ratio of the weight average (M_w) and number average (M_n) of the molecular weight [38, 39]. For a given M_n, high polydispersity (i.e., wide distribution) in an oligomer leads to high viscosity. This can be explained by the fact that the high molecular components of an oligomer have a significantly greater impact on viscosity than the low molecular components.

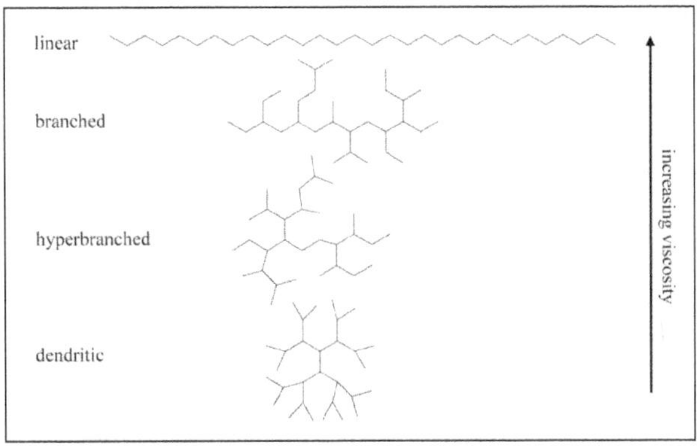

Figure 56: Viscosity is affected by the structure of the oligomer

The degree of branching of the oligomers also influences viscosity [40]. This is not necessarily the same as the functionality, because not every branched chain of a monomer or oligomer actually carries a polymerizable group. For a given molecular weight, the molecular volume decreases as the degree of branching increases, resulting in a low viscosity [41]. However, two cases must be distinguished. In the first and most common case, the oligomers possess a rather arbitrary distribution of a greater or lesser number of long, branched chains. This often results in a very wide molecular weight distribution and thus in a high viscosity (see above). In the second case, the oligomers are hyperbranched or dendritic, with ellipsoidal or spherical morphology [42]. These molecules have a low hydrodynamic volume in relation to their molecular weight, resulting in lower viscosities. Unfortunately, these types of oligomer tend to be rather expensive. Figure 56 shows the various structures for the same molecular weight.

Viscosity also increases with increasing intermolecular interaction between the molecules. Hydroxyl, carboxyl, ester, and urethane groups, for example, can form hydrogen bonds with one another. The viscosity therefore increases with the number of such groups (cf. Figure 64).

Stiff and rigid segments in a molecule lead to high viscosity because the molecule is prevented from forming a compact coil. A measure of chain stiffness is provided by the glass transition temperature (T_g). Molecules that form films with high T_g have a relatively high molecular volume in solution, resulting in a high viscosity [38, 39]. T_g of binders is influenced by the starting materials used in their preparation. For example, cycloaliphatic or aromatic structural elements result in higher T_g values than linear, aliphatic components of the same molecular weight. T_g is also influenced by molecular weight, the degree of branching, and the ability to participate in intermolecular interactions.

Not the least important factors influencing viscosity are the solubility of the oligomers and the dissolving power of the monomers. Briefly expressed, solubility in a monomer ("solvent") is increased by (a) reducing the intermolecular interactions between the oligomers, and (b) increasing the degree of solvation of the oligomer by the monomer. In the case of poor solubility, gel-like structures may be formed as a consequence of incompatibility, resulting in a higher viscosity. Following the "like dissolves like" rule of thumb, the choice of monomers is determined by the polarity of the resin. A more advanced thermodynamic treatment of the subject is available in [40, 43].

3.3.1.3 *Influence of monomers and oligomers on the reactivity of the formulation*

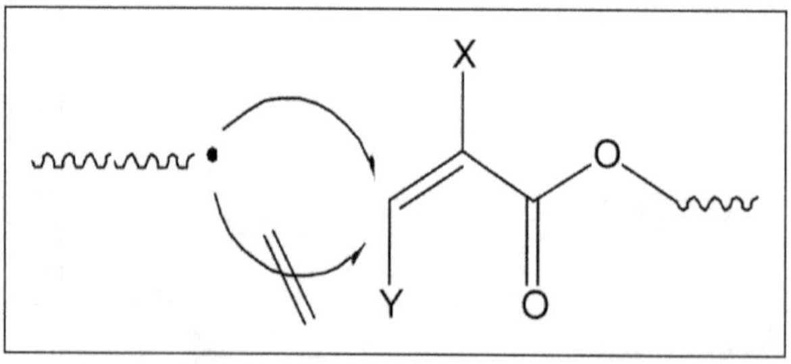

Figure 57: Steric influence of substituent Y on the double bond is greater than that of X during free radical polymerization

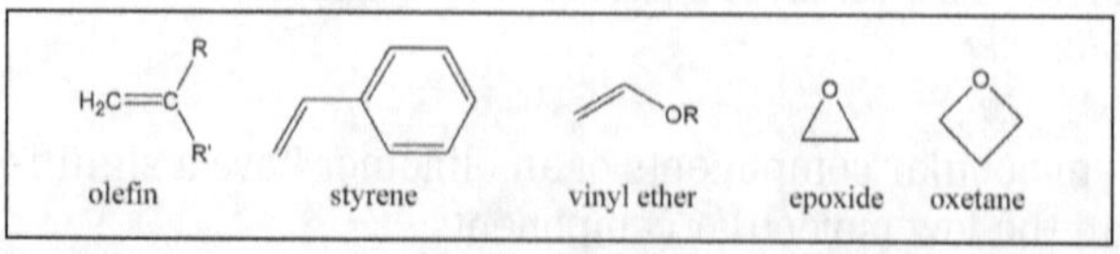

Figure 58: Structures that can be polymerized cationically

High throughput for curing coatings and inks can be achieved by selecting the right photoinitiators and radiation sources (see Chapter 3.2 and 4), as well as by the use of highly reactive monomers and oligomers. In practice, it is the macroscopic properties of the films that are important. For this reason, the development of mechanical or chemical resistance are usually investigated in order to describe the "reactivity" of a formulation. Typical test methods include the fingernail scratch and solvent tests (see Chapter 6). The test results are affected by the reactivity of the binders and the degree of their conversion. The reactivity is directly linked to the reaction rate, so that the rate and conversion of the polymerizable groups can also be investigated using techniques such as FT IR spectroscopy (see Chapter 6.2) [44].

At the molecular level, the chemical environment of radically curable double bonds influences their reactivity. As is shown in Figure 57, the monomer is attacked by the growing radical chain during polymerization. The steric requirements of substituent Y have a larger influence than those of X. Thus terminal acrylate double bonds are more reactive than double bonds lying along the oligomer backbone of the unsaturated polyesters. Additionally, the substituents X and Y have a polarization effect on the double bond. For example, the higher reactivity of acrylate functional monomers and oligomers as compared with their methacrylate functional analogs is ascribed mainly to the +I effect of the methyl group in the methacrylate; steric shielding by the methyl group plays only a subordinate role.

Similar considerations apply for cationic polymerization, but in this case the type of the polymerizable group must be considered. Electron-rich olefin derivatives and particularly cyclic ethers, are commonly used. The reactivity of electron-rich olefin derivatives increases with the strength of the +I effect of the substituents. In principle, electron-rich olefins and vinyl aromatic compounds like styrene can always be polymerized, but in practice vinyl ethers are most commonly used (Figure 58).

The cyclic ethers used are generally epoxides, and, less frequently, oxetanes. In this case electron-withdrawing groups reduce reactivity, which is the reason why propylene oxide groups are more reactive than glycidyl ethers [45]. Ring strain also influences reactivity: cycloaliphatic epoxides are more reactive than linear. The reactivity of cationically polymerizable groups decreases in the following order:

vinyl ether > propenyl ether > cycloaliphatic epoxide > propylene oxide > glycidyl ether.

The reactivity of vinyl ethers is at roughly the same level as that of radically curable acrylates [35].

Table 8: Glass transition temperatures (as determined by DSC) of homopolymers of selected monomeric acrylates with various functionalities [47] 1)

Monomer	T_g/°C
2-phenoxyethyl acrylate	5
isobornyl acrylate	88
isodecyl acrylate	-60
hexanediol diacrylate	43
tripropyleneglycol diacrylate	62
trimethylolpropane triacrylate	62
propoxylated trimethylolpropane triacrylate (PO_3)	-15
ethoxylated trimethylolpropane triacrylate (EO_3)	-40

1) www.sartomer.com/home.asp: *Glass Transition Temperatures of acrylate monomers*

Note

In the case of cationic photopolymerization, the reaction rate and degree of conversion can be simultaneously improved by adding hydroxyfunctional components, which are chemically incorporated into the film by chain transfer reactions. This also allows control of film properties. For example, the T_g of the film is decreased, resulting in improved flexibility (see Chapter 3.3.4.4).

Apart from the parameters described above, which are directly associated with the polymerizable groups, the reactivity of binders increases with the speed with which the monomers and oligomers move toward each other throughout the curing process. The crosslinking kinetics of radically curable systems is determined mainly by diffusion (*Norrish-Trommsdorff* effect). Due to the high (initial) viscosity, the number of chain termination reactions by recombination and disproportionation of polymer radicals is small. The simultaneous formation of additional radicals leads to an increase of the radical concentration, resulting in an accelerated reaction (see A in Figure 59) [46]. As the reaction progresses the glass transition temperature of the film being formed increases rapidly. In the case the T_g of the film reaches the ambient temperature, oligomer chains and radicals are „frozen" what reduces their mobility (vitrification) (see B in Figure 59). The rate of the subsequent reaction is therefore reduced, and full (100 %) conversion of all the double bonds is not achieved (see C in Figure 59). The T_g of the film depends directly on the T_g of the binders: a higher cure extent is achieved if the monomers and oligomers have a low T_g. Table 8 shows the homopolymer glass transition temperatures of some monomers.

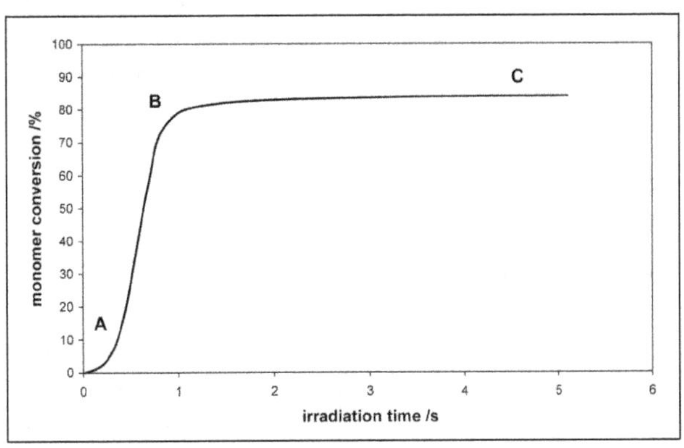

Figure 59: Polymerization behavior of acrylate monomers: Monomer conversion versus irradiation time

The T_g of the cured film depends also on its crosslink density, and therefore on the functionality of the components: the higher the functionality, the more rapidly the crosslink density increases during polymerization. Therefore, the T_g above which the radical chains are frozen is attained at a relatively early stage during conversion. Although the crosslink density raises when increasing functionality of the binders (cf. Figure 54), the *relative* conversion (number of polymerized double bonds/total number of double bonds) simultaneously diminishes. By elevating the film temperature by increasing the ambient or substrate temperature during curing, the reaction rate and degree of conversion can be improved [36, 48].

Note

The temperature effect may be reversed in the curing of thin films under air (< 5 μm) what decreases the film viscosity and hence favors oxygen diffusion through the film.

The free radical polymerization stops immediately when shutting the UV light source off due to vitrification or the fast consumption of the radical species by recombination. In contrast, cationic polymerization continues to proceed in the dark as long as the acid-initiator is not consumed (see Chapter 2.1.2). The degree of conversion increases with time, and films that initially had a good hardness-elasticity balance may become brittle.

Note

Monomers and oligomers are usually supplied with a small percentage of stabilizers, to increase storage stability by preventing unwanted polymerization (see Chapter 3.5.5). The amount of stabilizers in the total formulation should be as low as possible as they are likely to interfere with the polymerization process (usually <<500 ppm). It may be necessary to add additional amounts of photoinitiator to compensate their action. It should be noted that some stabilizers are toxicologically hazardous. For example, hydroquinone (1,4-dihydroxybenzene), which was commonly used, is possibly carcinogenic. It can be replaced by e.g., 2,6-di-tert-butyl-4-methylphenol [49].

As a summary, it must be emphasized that the reactivity of the formulation strongly depends on the selection of the monomers and oligomers. Steric factors, polarization of the polymerizable groups by substituents, and the ability of the substituents to participate in resonance stabilization all influence the reactivity. Additionally, a high glass transition temperature leads to a lower degree of conversion.

Table 9: Parameters influencing the adhesion of radiation curable formulations

Parameter	Suggested action
substrate wetting	surface tension of the formulation must be lower than that of the substrate; use a substrate wetting agent if required
adsorption between film and substrate	increase the number and strength of contacts to the substrate: use binders with hydrogen-bonding groups (urethane, OH, etc.) or acid groups in the case of metals (salt formation); use binders of low T_g; pretreat plastics if necessary.
diffusion into (absorbent) substrates	increase penetration by using a low-viscosity formulation (but see conversion below)
surface swelling of substrate (plastics)	use of suitable monomers with high dissolving power
conversion	increase conversion at the film/substrate interface if necessary; avoid excessively deep penetration into the substrate
volume shrinkage / internal stresses	use monomers and oligomers with the lowest possible functionality, highest possible molecular weight, high steric requirements, and low Tg; increase the temperature during the curing process

3.3.1.4 *Influence of monomers and oligomers on the chemical and mechanical properties of films*

As the major components in RC systems, monomers and oligomers decisively influence the mechanical and resistance properties of crosslinked films. They can be selected on the basis of their effects on adhesion, elasticity, hardness characteristics, and resistance to abrasion, chemicals, or weathering. These properties can be influenced by their molecular weight, functionality, and chemical structure. The homogeneity of the crosslinking reaction within the film and the degree of cure also plays a decisive role.

Influence of monomers and oligomers on adhesion

The adhesion of cured films to the substrate is influenced by a large number of parameters [38, 50, 51]. The most important parameters affecting adhesion are shown in Table 9.

The substrate must initially be wetted by the coating or ink as thoroughly as possible. This occurs when the surface tension of the formulation is lower than that of the substrate (see Chapter 3.5.3). The surface tension strongly depends on the polarity of the binders, which is essentially determined by the type and number of the functional groups. Polar groups, such as hydroxy or carboxy groups, increase the surface tension, while nonpolar groups, such as long alkyl chains, siloxanes, or (fluoro)alkyl groups, reduce it. For a more detailed discussion, see [52, 53].

Note

It is possible to reduce the surface tension of the formulation by adding long-chain aliphatic monomers such as octadecyl acrylate, which improves adhesion to a number of plastics [54, 55]. Surface tensions of some monomers are shown in Table 10. However, their use is restricted by compatibility, high volatility, low functionality, and, in some cases, high prices. Alternatively, special adhesion resins (e.g., special polyesters) or substrate wetting agents (see Chapter 3.5.3) can be used.

To reach a satisfying adhesion, the elasticity of the film must be adjusted to the substrate. A brittle film on a deformable substrate flakes off immediately. Additionally, the number of contacts per unit surface area between the film and the substrate, and the strength of these contacts (adsorption), are crucial in controlling adhesion. Mobile, flexible molecules (with low T_g) are more easily able to orient their potentially adsorbable groups toward the surface

Table 10: Volume shrinkage and surface tension of selected monomeric acrylates of various functionalities (F)

Name	Abbr.	F	Surface tension/mN/m (25 °C)	Volume shrinkage/%	Literature reference
isobornyl acrylate	IBOA	1	32	5.2	54
isodecyl acrylate	IDA	1	29	10.0	55
octyl/decyl acrylate	ODA	1	30	8.3	54
hexanediol diacrylate	HDDA	2	36	13.1/19.0	55/54
dipropylene glycol diacrylate	DPGDA	2	35	13.0	55/54
tripropylene glycol diacrylate	TPGDA	2	34	12.3/18.1	55/54
trimethylolpropane triacrylate	TMPTA	3	38	25.1	54

than those possessing stiff and rigid chains (high T_g). The strength of the adsorption naturally depends on the chemical character of the substrate and binder. In the case of metallic substrates, metal salts may be formed if the film-forming components contain carboxy or phosphoric acid groups. If the substrate contains polar groups, adsorption can be improved by dipole-dipole interactions (e.g., hydrogen bonds). For nonpolar substrates, e.g., a number of plastics like polypropylene, adsorption occurs only via weak dispersion forces (*van der Waals* or *London* forces), making it necessary to increase the polarity of the substrate by using appropriate pretreatment methods (e.g., primers, corona, or plasma pretreatment).

In the case the formulation can penetrate into the substrate, the number of contact points is strongly increased by "interlocking" the film and substrate. The most important factor here is the substrate, which can prevent penetration of the formulation (as in the case of metallic substrates) or can be absorbent (as in the case of wood or paper). In the case of absorbent substrates, a low viscosity achieved with the use of low molecular weight monomers and resins favors penetration. However, excessive, deep penetration prevents radiation from reaching the penetrated formulation during the curing process. The result is insufficient through-cure, leading to poor adhesion. Moreover, these uncrosslinked components can migrate through the substrate. In the case of food packaging, components of the coating system might then migrate into the food (see Chapter 6.2.3).

Note

Through-cure extent of the films can easily be verified by applying the formulation on release paper and curing under the appropriate conditions. The film is then stripped away from the substrate. If the film is greasy and matt on the side facing the substrate, the cure is not completely achieved: curing conditions must be optimized in order to improve adhesion (see Chapter 6.2).

Certain monomers can induce surface swelling of some plastics at the interface between film and substrate. This leads to the formation of an interpenetrating network between the film and substrate, markedly improving adhesion. A higher (substrate) temperature during crosslinking supports this process. Due to their low molecular weight and polarity, mono- and difunctional monomers are generally more effective in this regard than oligomers or higher-functional monomers. For example, dipropylene glycol diacrylate, hexanediol diacrylate, and lauryl acrylate cause surface swelling of plastics such as polycarbonate, PVC, and polystyrene [54, 55].

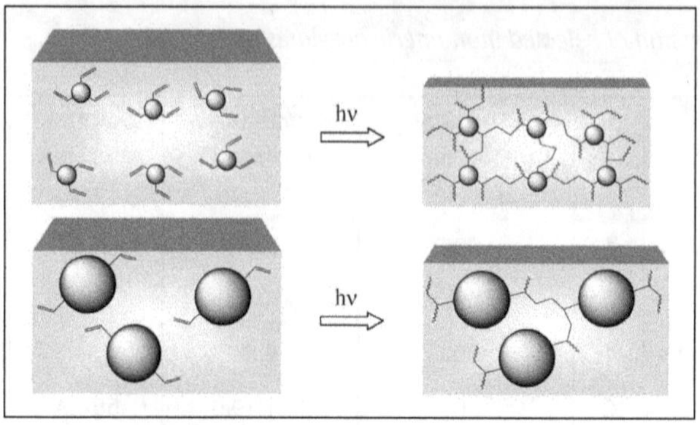

Figure 60: Model for volume shrinkage during the curing process: Above: Small molecules with high functionality lead to high volume shrinkage. Below: Volume shrinkage is reduced by the use of sterically demanding oligomers with low functionality

In radically curable formulations, the volume shrinkage that occurs during crosslinking leads to a significant reduction of the adhesion between substrate and film. This is explained by the fact that the distance between the acrylic molecules in the uncured state is higher than after conversion of the double bonds into covalent single bonds. High functionality, low molecular weight and low steric requirements of the film-forming substances lead to enhanced volume shrinkage. This effect is illustrated in Figure 60. Because monomers have a high functionality relative to their molecular weight, they have a significant influence on volume shrinkage. Table 10 gives functionalities and volume shrinkage for some monomers, while additional monomers are shown in Table 14.

Depending on the radiation source and the exposure time, temperature of the film during radiation lies between 40 and 80 °C. As long as the glass transition temperature of the film being formed lies below the curing temperature, tensions resulting from film formation can relax, so that the effect of volume shrinkage on adhesion is relatively slight. However, if the T_g of the film being formed rises above the curing temperature, molecules of the polymeric film are vitrified (i.e., in the glassy state) and internal stresses develop due to the abrupt onset of crosslinking and the associated volume shrinkage. This can lead to separation of the film from the substrate (delamination). It is therefore necessary to release the internal stresses within the film. This can be achieved by keeping the T_g and the crosslink density of the film as low as possible [51]. Frequently, these parameters cannot be adjusted because of the mechanical properties desired in the film. If the thermal expansion coefficients of the (heat resistant) substrate and film are similar, then internal stresses can be released by increasing the film temperature during the curing process by increasing the substrate or ambient temperature [48]. If, however, the thermal expansion coefficients markedly differ, additional internal stresses occur during cooling, which can lead to loss of adhesion [56].

One advantage of cationically curable formulations containing epoxides is the relatively low volume shrinkage and the resulting good adhesion to the substrate. This is due to the fact that the epoxide ring is opened during crosslinking, forming an ether bridge that shows approximately the same distance between the molecules before and after conversion.

Note

- Volume shrinkage and internal stresses can also be reduced by small quantities of saturated, non-crosslinkable binders. These adhesion resins improve adhesion, inter-layer adhesion, and flexibility [37]. However, larger quantities impair the resistance properties of the films.
- In some applications the free radical polymerization of the applied film is first initiated by a relatively low UV dose. The resulting film has a low crosslink density and can adapt its shape. After a second irradiation step, the film is fully crosslinked. Thus the effect of volume shrinkage on adhesion is relatively slight because the extended polymerization time reduces internal stresses.

Influence of monomers and oligomers on the resistance properties

Because the crosslink density is generally higher for radiation cured films than for conventionally cured films, properties such as hardness, scratch resistance, and resistance to heat, water, and chemicals are generally excellent. For an acrylic molecule of a given basic structure, the crosslink density augments when increasing functionality and decreasing molecular weight (Figure 54). Monomers and oligomers that produce films with high T_g (see Chapter 3.3.1.2) also have a positive influence on the resistance properties of the film. However, the influence of T_g on the degree of conversion must be taken into account (see Chapter 3.3.1.3). At the same time, the antiblocking properties can be improved by achieving a high T_g resulting in a better efficiency of, e.g., stacking piece products soon after application and curing.

However, high crosslink densities and T_gs can have adverse effects on elasticity and adhesion. Therefore, selection of monomers and oligomers has to be realized in order to find the best compromise between the above-mentioned properties depending on the application.

Within certain limits, good balance between hardness and elasticity can be attained by the use of higher molecular oligomers of functionality between two and four. Hydrogen bonds also have positive effect on the hardness-elasticity balance. This is one of the reasons why urethane acrylates have good mechanical properties.

Note

A high glass transition temperature can also be achieved by adding non-curable hard resins, which simultaneously reduce volume shrinkage (through reduction of the cross-link density) and internal stresses (see Chapter 3.3.2.11). An example is provided by ketone-aldehyde resins [57].

Influence of monomers and oligomers on weathering and yellowing resistance

To obtain good resistance to weathering and yellowing, the same basic principles apply for RC coatings as for conventional systems. (Cyclo)aliphatic groups are always to be preferred to aromatic groups.

The resistance of polyether acrylates to oxidation processes is relatively low. Aliphatic polyester acrylates have good oxidation resistance, but may have lower resistance to hydrolysis. Aliphatic urethane acrylates have the best weathering and yellowing resistance of all oligomers. A high crosslink density helps retard visible polymer degradation [58].

In the case of cationically curable systems, vinyl ethers tend to yellow more than epoxides and are less stable to weathering (see Chapter 3.3.4.2). Table 11 is intended to facilitate the selection of raw materials by summarizing the basic characteristics of monomers and oligomers and their effects on coatings and inks.

Table 11: General structure-property relationships of monomers and oligomers

Molecular weight: M_n, M_w

Definition	Characteristics	Effect: high molecular weight leads to
measure of the chain length of the monomers and oligomers	M_n: number average M_w: weight average	• high viscosity • low reactivity • low crosslink density • poor solubility • less intense odor • low migratability • low volume shrinkage (better adhesion) • a large number of contact points to the substrate (better adhesion) • low penetration and low capacity to swell the substrate (poorer adhesion on plastics) • good hardness-elasticity balance • good resistance to chemicals

Table 11: General structure-property relationships of monomers and oligomers (continued)

Molecular weight distribution, polydispersity (PD)

Definition	Characteristics	Effect: broad molecular-weight distribution leads to
measure of the non-uniformity of molecular weight polydispersity (PD) = M_w/M_n	In the case of a broad molecular-weight distribution, the oligomer has a number of high-molecular as well as low-molecular components. The PD increases when using branched monomers in the oligomer synthesis.	• high viscosity • low tendency to crystallization

Degree of branching

Definition	Characteristics	Effect: a high degree of branching leads to
measure of the number of branches in an oligomer	A high degree of branching increases steric requirements, and is achieved by the use of branched monomers in the synthesis of the oligomers.	• broad molecular-weight distribution (in general) • good solubility and compatibility of the oligomers (short-chain branching) • low volume shrinkage

Functionality, F

Definition	Characteristics	Effect: high functionality leads to
average number of polymerizable groups per molecule	The higher the functionality, the greater the number of polymerizable groups carried by the monomer or oligomer.	• high reactivity • high crosslink density • high volume shrinkage (poor adhesion) • low content of migratable components

Crosslink density

Definition	Characteristics	Effect: high crosslink density leads to
number of network points per unit volume	High functionality and low molecular weight of the monomers and oligomers, and a high degree of conversion, increase crosslink density.	• high T_g • high hardness, high scratch and chemical resistance • low elasticity and flexibility

Type of the polymerizable groups/1

Definition	Characteristics	Effect: acrylate functional systems are
acrylate vs. methacrylate	Both groups polymerize radically. Steric requirements and T_g are lower for acrylates.	• more reactive • more flexible • less voluminous (volume shrinkage, adhesion) as compared with methacrylate

Table 11: General structure-property relationships of monomers and oligomers (continued)

Type of the polymerizable groups /2

Definition	Characteristics	Effect: epoxy groups give rise to binders with
epoxide vs. vinyl ethers	Epoxide groups crosslink via ring-opening, and vinyl ethers via "normal" cationic polymerization. Vinyl ethers can also be radically cured.	• lower reactivity • lower volume shrinkage • higher resistance to yellowing and heat as compared with vinyl ethers

Glass transition temperature (T_g) of monomers and oligomers

Definition	Characteristics	Effect: high T_g of the non-cured monomers or oligomers leads to
temperature below which the homopolymerized monomers and oligomers are in the glassy state	Monomers and oligomers possess a high T_g if they are rigid and stiff, can form hydrogen-bonds, and have high functionality.	• high viscosity • low reactivity of the liquid formulation

Glass transition temperature (T_g) of the crosslinked film

Definition	Characteristics	Effect: high T_g of the cured film leads to
temperature below which the crosslinked components are in the glassy state	High T_g is obtained by using rigid, stiff monomers and oligomers, hydrogen-bonding components, and a high crosslink density.	• low flexibility/elasticity • high hardness and scratch resistance • good antiblocking properties • good chemical and water resistance • good barrier properties against chemicals, water, and oxygen • low degree of through-cure

Steric requirements

Definition	Characteristics	Effect: high steric requirements lead to
measure of the molecular volume of the film-forming molecules	Use of bulky monomers (educts) for synthesis of the monomers and oligomers and a high degree of branching increase the steric requirements.	• low mobility, reactivity, and degree of through-cure • low volume shrinkage (better adhesion) • low surface swelling behavior

Aliphatic groups

Definition	Characteristics	Effect: a high number of aliphatic groups leads to
introduction of linear saturated hydrocarbon groups	Linear groups lead to a low T_g and have low steric requirements.	• high reactivity • low viscosity • high volume shrinkage • high flexibility • good weathering stability

Table 11: General structure-property relationships of monomers and oligomers (continued)

Cycloaliphatic groups

Definition	Characteristics	Effect: a high number of cycloaliphatic groups leads to
introduction of cyclic saturated hydrocarbon groups	Cycloaliphatic groups result in a high T_g and have high steric requirements.	• high viscosity • low volume shrinkage • good resistance properties (heat and scratch resistance, weathering stability) • good hardness-elasticity balance

Aromatic groups

Definition	Characteristics	Effect: a high number of aromatic groups leads to
introduction of aromatic groups	Aromatic groups lead to high T_g and have high steric requirements. Light absorption is also high.	• low volume shrinkage • low flexibility • good hardness and scratch resistance • poor weathering resistance, particularly of clear coatings

Polar functional groups

Definition	Characteristics	Effect: a high number of polar groups leads to
introduction of additional functional groups (-OH, -COOH, -NH, ester, urethane) (see Chapter 6.1.1)	Polar groups lead to formation of intermolecular hydrogen bonds. In cationically curable systems, polar groups affect the crosslinking reaction (e.g., inhibition of superacids by basic groups, chain transfer by OH groups)	• high viscosity • high solubility of the oligomers in polar monomers • less intense odor • high T_g • high surface tension (substrate wetting) • good adhesion to polar substrates • poor water resistance (swelling behavior of, e.g., polyethers) • good mechanical properties (e.g., urethane groups)

Chain transfer, CT

Definition	Characteristics	Effect: chain transfer leads to
chain transfer agents reduce the molecular weight during crosslinking	Molecular weight is reduced without any effect on the reaction rate or degree of cure. Examples of CT agents: radical: thiols cationic: alcohols	• low crosslink density (high flexibility, low hardness). In cationically curable systems, alcohols increase the degree of through-cure.

3.3.2 *Oligomers for curing by free-radical polymerization*

Radically polymerizable oligomers usually bear acrylate groups. Methacrylates have low reactivity and are therefore used only in specific cases (see Chapter 3.3.1.2 and 3.3.1.3). Cyclic vinyl amides or vinyl ethers are sometimes used in combination with acrylate functional components to reduce the viscosity of the formulation (see Chapter 3.3.4.2).

The toxicological effect (e.g., skin or eye irritation) of oligomers is generally less pronounced than for monomers, as a result of their higher molecular weight. Contact with the components should nevertheless be avoided: long and repeated exposure may sensitize the skin (see Chapter 7).

3.3.2.1 *Epoxy acrylates*

Due to their low price and high reactivity epoxy acrylates are widely used in radiation curable formulations. This class of compounds is derived from bisphenol A epoxides, epoxy novolacs, and epoxidized oils. Oligomers based on bisphenol A diglycidyl ethers are predominantly used. In general, epoxy acrylates have a high solution viscosity as a result of their molecular weight and the large number of hydrogen bonds arising from the presence of hydroxy groups. The reactivity is high even when curing is performed under air. The reason lies in the β-hydroxy ester groups present: abstraction of the hydrogen at the carbon atom carrying the hydroxy group results in species that can initiate polymerization and react with atmospheric oxygen to form reactive radicals (Figure 61) [59].

Bisphenol A epoxy acrylates

Chemical structure and preparation

Bisphenol A epoxy acrylates are obtained by reacting epoxy functional oligomers with acrylic acid in a ring-opening addition reaction under mild conditions (Figure 62). In specific cases, the hydroxyl groups formed are reacted further with acrylic acid derivatives to give higher-functional epoxy acrylates. Properties of the oligomers can be modified by varying the molecular weight of the starting bisphenol A epoxy resins; bisphenol A diglycidyl ether is essentially used here because higher homologs are solids. Other bisphenols such as bisphenol F are used to a less extent as starting materials. Modification with fatty acids reduces the polarity and viscosity of the epoxy acrylates; the flexibility and pigment-wetting behavior are also improved [60].

Properties

As stated above, the viscosity of bisphenol A epoxy acrylates is high; the molecular weight is therefore set at a low level in order to obtain good flow behavior and allow an easy application. Nevertheless, reactive diluents are usually used to reduce viscosity. The reacti-

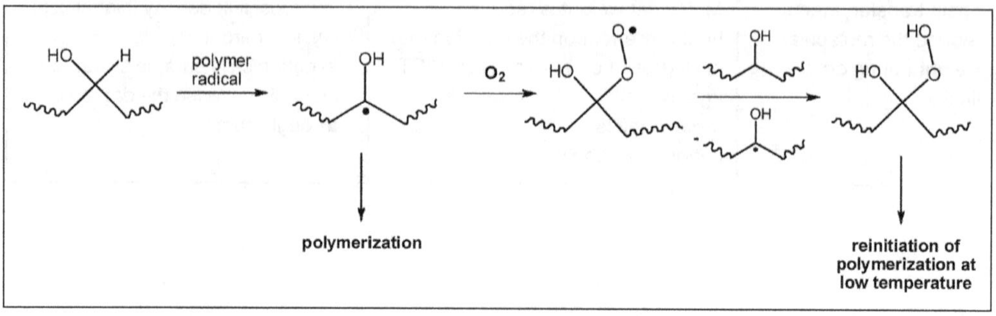

Figure 61: Epoxy acrylate radicals can be formed by abstraction of hydrogen atoms [59]

vity of the oligomers is generally very high, even in the presence of atmospheric oxygen (see above). The resulting film exhibits a high gloss, high hardness as well as good chemical and corrosion resistance. However, yellowing and weathering resistance of the aromatic systems are relatively poor, as is the flexibility of the films on account of their high hardness. Moreover, pigment wetting is poor in the unmodified varieties.

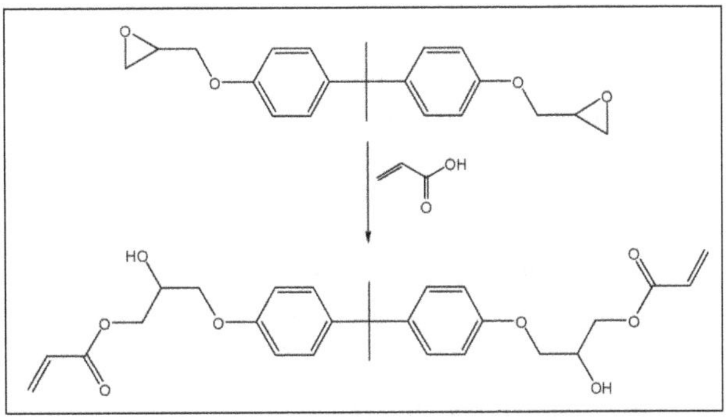

Figure 62: Difunctional bisphenol A epoxy acrylate oligomers can be prepared from bisphenol A diglycidyl ether and acrylic acid

Areas of application

Epoxy acrylates occupy a major position in the oligomer market and are used for UV and EB curable coatings for wood (e.g., parquet), metal, plastics, overprint varnishes, wet offset lithographic inks, and silk-screen inks for paper and cardboard. They allow a satisfying compromise between material performance and cost of raw material.

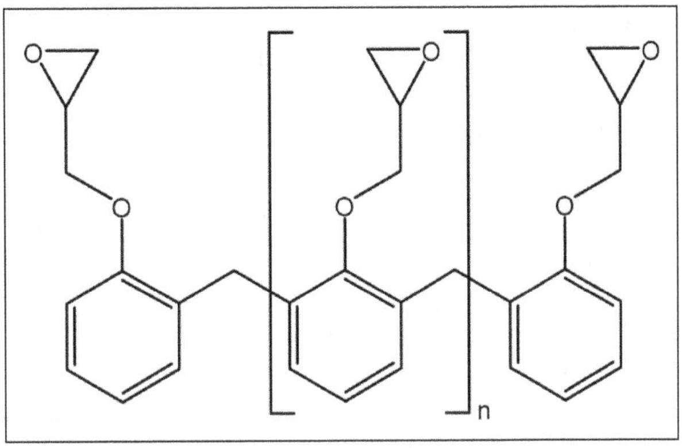

Figure 63: Chemical structure of an epoxy novolac

Epoxy novolac acrylates

Chemical structure and preparation

Epoxy novolac acrylates are prepared similarly to bisphenol A epoxy acrylates, by reacting the epoxide ring with acrylic acid. The starting materials are epoxy novolacs (Figure 63). As with other oligomers, the properties can be adjusted by varying the molecular weight of the starting resins.

Properties

Reactivity and functionality are usually very high. The resulting films have a high crosslink density and are extremely hard and resistant to chemicals and heat. These properties are supported by a high T_g. However, the cured films are brittle and inflexible (with low impact resistance and elongation at break, and high tensile strength). The aromatic groups lead to poor weathering resistance. Monomers are usually employed to reduce the viscosity.

Areas of application

Epoxy novolac acrylates are specialty products used to coat hard, inflexible substrates such as hardwood, metal, and glass. Due to their excellent heat resistance and high reactivity, they are used in the electronics industry as solder resist to cover soldered joints. They are not suitable for coating of flexible substrates.

Acrylates of epoxidized oils

This group of oligomers is derived mainly from epoxidized oils such as soy oils. They are synthesized analogously to the bisphenol A epoxy acrylates.

Properties

Acrylated oils usually have low viscosity, low reactivity, and good pigment wetting properties. The final film is flexible with low mechanical resistance. Adhesion to a number of substrates is outstanding [60].

Areas of application

Acrylated oils are usually combined with other oligomers to improve the above mentioned properties of coatings and printing inks, and also to reduce costs. They are commonly used in radiation curable printing inks and overprint varnishes, e.g., for paper.

Other acrylates of epoxidized products

Amine modified epoxy acrylates are also used in radiation curing, as the amino group in combination with photoinitiators accelerates the crosslinking reaction (see Chapter 3.3.2.4). Adhesion to plastics is an added benefit. Other specialty products are based on, for example, linear, epoxidized polybutadiene oils, which are used to enhance flexibility and improve adhesion on certain critical substrates.

3.3.2.2 Urethane acrylates

Urethane acrylates have outstanding properties. Of all the oligomers, they possess the best balance between hardness, toughness, and chemical and abrasion resistance on the one hand, and flexibility and elasticity on the other. Possibilities for variation are generally wide, so that extremely hard and abrasion-resistant films can be produced as well as extremely soft and elastic ones. Many of these properties are based on the ability of urethane groups to form intermolecular hydrogen bonds, which, however, simultaneously increases the viscosity of the formulation (Figure 64) [61]. If the properties of the films were the sole consideration, urethane acrylates would undoubtedly be the most frequently used oligomers in radiation curable formulations. Apart from their high viscosity, the comparatively high price is a deterrent to their being more widely used.

Chemical structure and preparation

Simple urethane acrylates can be obtained by reacting di- or polyfunctional isocyanates with hydroxyalkyl acrylate components. Highly sophisticated urethane acrylates are produced by the more complex reaction of a polyol with diisocyanates and hydroxyalkyl acrylates. Two different preparative routes exist here:

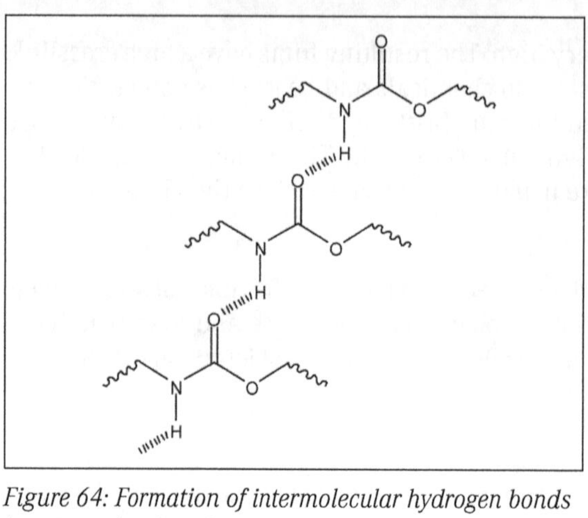

Figure 64: Formation of intermolecular hydrogen bonds increases the viscosity

1. Diisocyanate is first reacted with the hydroxyalkyl acrylate in such a way that one isocyanate group remains intact; this adduct is then further reacted with the polyol to form a urethane acrylate (Figure 65). This procedure yields well defined products; adducts formed from one mole of diisocyanate and two moles of hydroxyalkyl acrylate may be by-products of the reaction. Monomeric hydroxyalkyl acrylate residues are avoided.
2. Diisocyanate is reacted with the polyol to form an isocyanate-containing prepolymer. Higher

molecular products are also produced by reaction of two polyol molecules with one of diisocyanate; while this may be desirable, it greatly increases the viscosity. In the second step, the prepolymer is reacted with the hydroxyalkyl acrylate to form the urethane acrylate. In this synthetic route, monomeric hydroxyalkyl acrylate usually remains in the system which is a concern regarding hygiene.

If only diols are used and the reaction engineering is appropriate, the structural differences between the products obtained by the first and second synthetic routes are small. However, the second route yields polyurethanes with a wider molecular weight distribution and higher viscosity, especially in the case of higher functional polyols.

Hydroxyalkyl acrylates normally used are 2-hydroxyethyl acrylate (HEA) and 2-hydroxypropyl acrylate (HPA), which can be obtained by reacting ethylene oxide or propylene oxide with acrylic acid. Less frequently, hydroxybutyl acrylate is used. Because the HPA used is a mixture of molecules containing primary as well as secondary hydroxyl groups, adducts with diisocyanates are less homogeneous. This, and the additional methyl group, result in a lower tendency toward crystallization as compared with the HEA adducts [61]. The acrylate functionalities of the urethane acrylates vary between one and six, but usually lie between two and three.

Polyols can be simple diols such as 1,6-hexanediol. However, the urethane acrylates are often prepared from oligoester or oligoether polyols. Possibilities for variation, particularly for the oligoesters, are virtually unlimited due to the large number of available monomers and the regulation of, for example, degree of branching and molecular weight. Compared with polyether-based urethane acrylates, the more expensive oligoester-based urethane acrylates have higher reactivity, hardness, and solvent resistance, and a lower tendency to yellowing. They are, however, less resistant to hydrolysis, less flexible, and more viscous.

Aliphatic urethane acrylates

This class of oligomers produces films of the highest quality. Properties such as adhesion, hardness, and abrasion resistance, combined with high elasticity and flexibility, are excellent, as is resistance to weathering and yellowing. As a result of these outstanding properties, approx. 80 % of the urethane acrylates are based on aliphatics. The preparation of such products follows the routes described above. Two types of isocyanates are most commonly used: cycloaliphatic isophorone diisocyanate (IPDI) and dicyclohexylmethane-4,4'-diisocya-

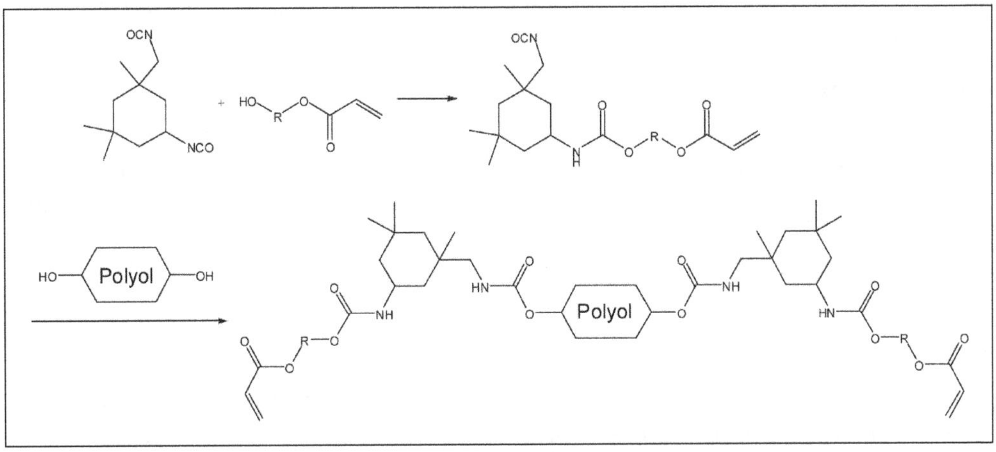

Figure 65: Preparation of a urethane acrylate by route 1 from isophorone diisocyanate, a hydroxyalkyl acrylate (where R is, e.g., C_2H_4 or C_3H_6), and a polyol (e.g., polyester polyol) leads to well defined products and only small residues of free hydroxyalkyl acrylate

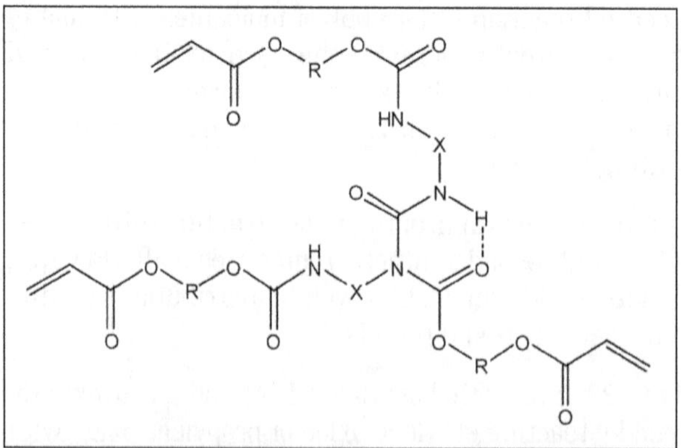

Figure 66: Schematic structure of HDI allophanate urethane acrylate, X = -(CH$_2$)$_6$-, R = e.g., -(CH$_2$)$_2$- [61, 63]

nate ($H_{12}MDI$). The latter leads to higher viscosities due to lack of selectivity. In contrast, the different reactivities of the two isocyanate groups in IPDI allow the synthesis of well defined products (Figure 65). These isocyanates provide good hardness and tensile strength but relatively low flexibility (elongation at break), which can usually be improved by appropriate selection of polyols. Linear hexamethylene-1,6-diisocyanate (HDI) is rarely suitable for flexibilization because the HDI hydroxyalkyl acrylate adducts tend to crystallize and have poor compatibility with other oligomers [26, 62]; HDI is nevertheless used for preparation of oligomers. In contrast, oligomers from trimethylhexamethylene diisocyanate (TMDI) have a low tendency to crystallize and a low viscosity. TMDI-based oligomers produce non-tacky and resistant films of high flexibility. TMDI flexibilizes IPDI-based oligomers without negatively impacting compatibility [62].

Higher-functional isocyanates such as IPDI-isocyanurate and HDI-isocyanurate increase the functionality of the oligomers. Due to their high viscosity they are currently used only to a limited extent for adjustment of film properties. In order to reduce the viscosity, HDI allophanate urethane acrylates have been developed (Figure 66) [61, 63].

Areas of application

Aliphatic urethane acrylates are suitable for high-grade exterior applications due to their resistance to weathering and yellowing. They are used when the above mentioned properties need to be of a very high standard. Due to their high viscosity, a compromise has to be made between viscosity and film properties; for this reason they are often combined with urethane-free oligomers or monomers [61]. They can be applied to all conceivable substrates such as plastics, metal, wood, paper, textiles, and leather. The main applications are industrial coatings for plastics and metals as well as high-quality furniture and flooring (parquet, PVC, or cork flooring). In the area of printing inks, they are often used for overprint varnishes and offset and screen inks.

Aromatic urethane acrylates

Aromatic urethane acrylates usually have poorer resistance to weathering and yellowing, higher viscosities, and lower flexibility than their aliphatic counterparts and are therefore less often used. Their properties and prices lie between those of polyester acrylates and aliphatic urethane acrylates. They are prepared similarly to the aliphatic systems with toluene diisocyanate (TDI) and methylene diphenylisocyanate (MDI) as the most commonly used aromatic isocyanates.

Areas of application

The applications of aromatic urethane acrylates are similar to those of their aliphatic counterparts, but are restricted to areas in which weathering and yellowing resistance are not of prime importance. Examples are coating of hardwood (e.g., parquet flooring), furniture, textiles, and leather, as well as overprint varnishes and offset, flexo, and screen inks. In the electronics industry, they are used in solder resists for printed circuit boards.

3.3.2.3 *Polyester acrylates*

Chemical structure and preparation

The most cost-effective method of preparing polyester acrylates is by esterification of the OH-groups of polyester polyols with acrylic acid or its derivatives (see Figure 55). Functionalization is also possible by reaction of the carboxyl groups of acid polyesters with hydroxyalkyl acrylates or glycidyl methacrylates. The starting materials used are linear polyester diols prepared from diols, dicarboxylic acids, and polyester polyols, that are branched to a greater or lesser degree, which additionally contain higher-functional raw materials such as trimethylol propane. Aliphatic monomers such as adipic acid, neopentyl glycol, and hexanediol are usually used to improve flexibility and weathering resistance. Cycloaliphatic groups (e.g., 1,2-cyclohexanedicarboxylic acid or 1,4-cyclohexanedimethanol) are used to improve adhesion (volume shrinkage) and mechanical properties. The more cost-effective aromatic monomers, such as phthalic anhydride and isophthalic acid, are used when weathering and yellowing resistance are not of major importance. To improve the wetting of organic pigments, the polyesters can be modified by fatty acids. The commonly used polyester acrylates have functionalities between two and six.

Properties

By varying the monomers, the degree of branching, and the molecular weight of the polyesters, it is possible to customize the glass transition temperature (to give hard or flexible resins), viscosity, functionality, reactivity, and many other properties for a particular area of application. Apart from the polymerizable group and the ester function, no other functional groups are present; this means that, for a given molecular weight, the viscosities of the polyester acrylates are low as compared with epoxy and urethane acrylates, so that they can sometimes be used even without additional monomers. Compatibility with other oligomers is high, as is resistance to yellowing. Due to the ester group's sensitivity to hydrolysis, the saponification stability is lower than for epoxy and urethane acrylates, but is nevertheless adequate for most purposes. Performance characteristics lie between those of epoxy and urethane acrylates.

Areas of application

Due to the highly favorable cost-performance ratio and wide possibilities of variation, polyester acrylates are well suited for a variety of applications. They provide the right balance of properties in an economical fashion. They are suitable for UV overprint varnishes for paper and cardboard, printing inks (offset, flexographic and inkjet inks and wet lithography) and wood coatings (furniture lacquers and parquet flooring). Due to their low viscosity and high flexibilizing action they are commonly combined with other oligomers. Fatty-acid-modified polyester acrylates are used for cost-effective improvement of pigment wetting, mainly in printing inks. Table 12 shows some properties of the three most important classes of oligomers.

3.3.2.4 *Polyether acrylates*

The range of possibilities for variation in the preparation of polyether acrylates is less wide than for polyester acrylates, on account of the limited number of raw materials. The monomers used are usually ethylene oxide and propylene oxide, which are polymerized with, for example, mono-, bi- or higher hydroxy functional starter molecules such as neopentyl glycol or trimethylol propane. The hydroxy functional polyethers obtained are acrylated by, for example, the route shown in Figure 55. The polyether acrylates exhibit binder or monomer character, depending on the degree of alkoxylation. The move from monomer to binder is fluid and a matter of definition.

Table 12: Important classes of oligomers and their basic properties [64]

Property	Epoxy acrylate (based on bisphenol A)	Polyester acrylate	Urethane acrylate
viscosity	medium to high	low to medium	medium to high
reactivity	very good	good	good to moderate
hardness/scratch resistance	very good	good to moderate	good
toughness	moderate	moderate to poor	very good
abrasion resistance	moderate to poor	good	very good
flexibility/impact resistance	moderate to poor	good to moderate	very good
weathering and yellowing resistance	moderate to poor	good to moderate	very good to good (aliphatic)
water resistance	very good	good to moderate	very good
chemical resistance	very good	good	very good
adhesion	good to moderate	good to moderate	very good

Polyether acrylates are highly flexible (with low T_g). They have the lowest viscosity of all acrylate oligomers (usually less than 100 mPa·s at 25 °C), so that only a small proportion of monomer, or none at all, is required to lower the viscosity. Many members of this class are water-soluble, which negatively impacts the resistance of the film to water and chemicals.

Polyether acrylates have high reactivity. Because the hydrogens at the carbon atom in the α-position to the ether oxygen are abstracted during radical curing, reactivity and crosslink density can be further increased by the use of amine synergists (cf. Figure 61). This has a positive effect on (surface) curing, and offers the possibility of targeted improvement of water resistance. On the negative side, this is the reason for the poor weathering and yellowing resistance, which is further aggravated by the presence of amines.

Areas of application

Instead of volatile, odorous monomers, polyether acrylates are often used in formulations in combination with other oligomers to reduce viscosity and volume shrinkage. This allows development of monomer-free systems. The main areas of application include plastic coatings, wood coatings for furniture, and paper coating. Many varieties are soluble or dispersible in water, and can therefore also be used in water-based systems.

3.3.2.5 Amine modified polyether acrylates

Amine modified polyether acrylates are prepared by the reaction of a small fraction (approx. 2-15 %) of the acrylate double bonds of a polyether acrylate with an amine in a Michael addition (Figure 67).

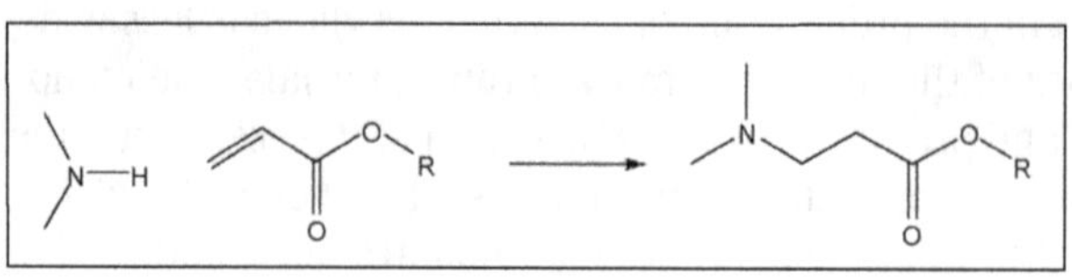

Figure 67: Amine modified polyether acrylates can be prepared by Michael addition of a secondary amine to the double bond of a polyether acrylate (R = polyether acrylate)

Amine modified polyether acrylates are used in small quantities in combination with Type II photoinitiators to increase the reactivity of formulations. They act here as co-initiators and reduce the concentration of atmospheric oxygen in the film (Figure 68).

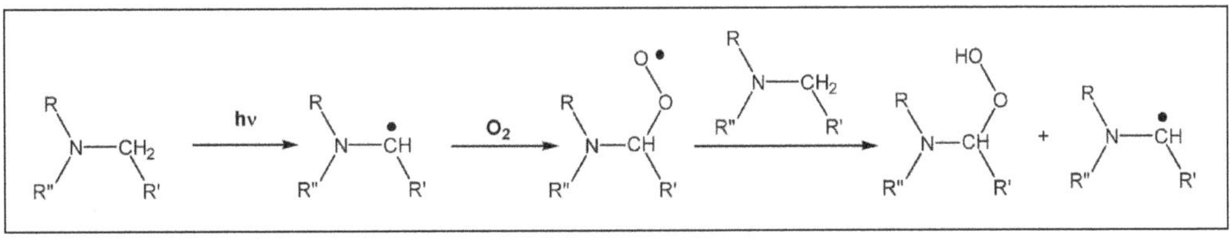

Figure 68: Amine modified polyether acrylates accelerate curing in the presence of atmospheric oxygen by reducing the oxygen concentration (R = polyether acrylate; R' = H, alkyl; R'' = alkyl)

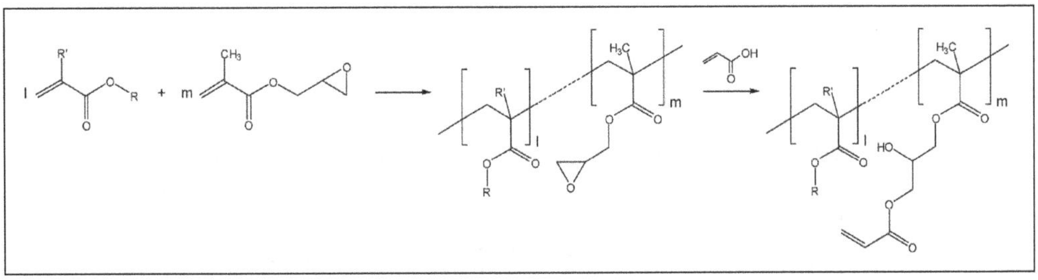

Figure 69: Acrylated oligo(meth)acrylates can be obtained by polymerization of (meth)acrylate monomers and GMA and subsequent reaction of the epoxy group with acrylic acid (R = e.g., alkyl; R' = H, CH_3)

This enables the formulator to avoid volatile low-molecular amines, thereby reducing odor and extractables [65, 66]. Because their viscosities are low, they reduce the viscosity of the total formulation. Reactivity is generally very high, and hardness and resistance to water and chemicals are significantly better than for polyether acrylates. The yellowing resistance is, however, poorer (see Chapter 3.2).

Areas of application

Amine modified polyether acrylates are used mainly as amine synergists (co-initiators) in combination with Type II photoinitiators in UV curable systems where high reactivity is desired and yellowing is not a prime concern. The main areas of application include printing inks, wood coatings for furniture, and paper coating.

3.3.2.6 Acrylated oligoacrylates

In contrast to the oligomers considered so far, in which the double bonds are generally located at the end of the oligomer backbone, the acrylate function in acrylated oligo(meth)acrylates is more or less statistically distributed over the side chains of the oligomer. Low-molecular oligo(meth)acrylates are synthesized e.g., by using glycidyl (meth)acrylate (GMA) as comonomer in a first step. The epoxy group is subsequently reacted with acrylic acid (Figure 69).

Thanks to the variety of acrylate and methacrylate monomers available, the possibilities for variation of the properties are very wide. The most important properties of this type of binder are very good weathering, chemical, and yellowing resistance. The use of the acrylated oligoacrylates is now restricted to certain niche applications, such as support of adhesion on problematic substrates and some exterior applications [37].

3.3.2.7 Silicone acrylates

The basic structure of silicone acrylates often derives from polydimethylsiloxanes (Figure 70). They can be prepared by hydroxyalkylation of polydimethylsiloxanes containing silane groups, followed by esterification of the hydroxyl groups with acrylic acid (variant I, Figure 71). Alternatively, epoxy groups can first be introduced, which are then reacted with acrylic

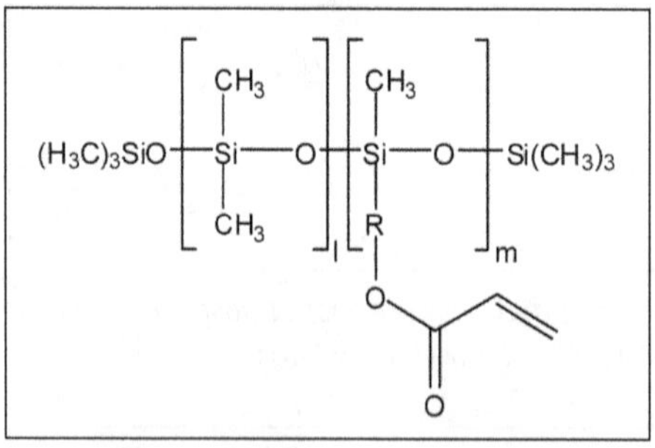

Figure 70: Chemical structure of a silicone acrylate based on a polydimethylsiloxane (R = alkyl)

acid. The ring opening results in the formation of hydroxyl groups in addition to the acrylate functionality (variant II, Figure 71) [67, 68].

A further preparative option consists in the hydrosilylation of multifunctional organic acrylates. This provides direct access to radiation curable silicones with good release properties (Figure 72).

The silicone backbone provides the products high flexibility and good resistance to heat, chemicals, and weathering. Silicone acrylates exhibit a strong hydrophobing action, which makes the coating films dirt and water repellent. The alkyl groups provide the necessary compatibility with the other components of the formulation. The crosslinking reaction is strongly inhibited by oxygen: in the past, curing had to be carried out under strictly oxygen-free conditions under an inert atmosphere. However, significantly higher concentrations of oxygen can now be tolerated with the use of amine synergists and deaeration additives (see Chapter 3.2.2 and 3.5). The development of epoxy functional silicones for cationic curing allows curing to be carried out even under normal atmospheric conditions [67, 68].

Areas of application

UV curable silicone acrylates can be applied as release coatings on paper and plastic substrates. Typical applications are release coatings for self adhesive label stock, which account for approx. 70 % of overall consumption (see Chapter 5.7.1). Other major fields of application include tapes and personal hygiene products like seals for diaper tapes. The formulations exhibit a high release effect over a long period of time. Specialty silicone acrylates are used as crosslinkable additives in radiation curable systems to improve properties such as substrate wetting, flow, and slip (see Chapter 3.5.6).

3.3.2.8 Self-initiating acrylate resins

The acrylate oligomers described above crosslink on irradiation if a photoinitiator is present that initiates the polymerization. A new kind of oligomer has been introduced recently that actually does not need any additional photoinitiators [69, 70, 71]. These self-initiating, UV curable acrylate resins are obtained by a Michael reaction between β-keto esters and diacrylates such as monomeric acrylates or epoxy, urethane, or polyester acrylates (Figure 73). Upon

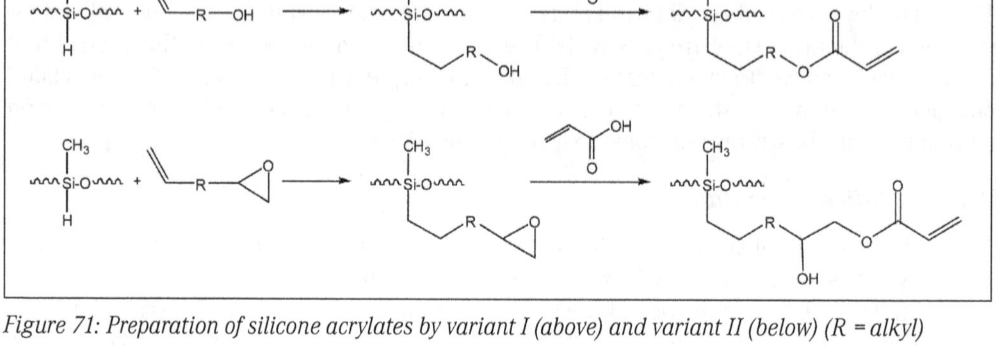

Figure 71: Preparation of silicone acrylates by variant I (above) and variant II (below) (R = alkyl)

irradiation radicals are formed that initiate the polymerization (Figure 74). Due to the absence of low molecular weight photoinitiators the level of extractables in cured films is low. High viscosity remains a key issue so that reactive diluents and small amounts of photoinitiators are in fact additionally required to allow an easy handling of the formulation.

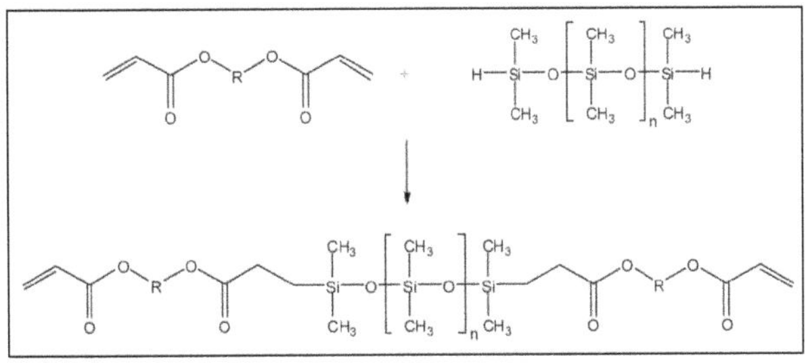

Figure 72: Preparation of silicone acrylates by direct hydrosilylation (R = alkyl)

<u>Areas of application</u>

Due to a low level of extractables in the cured film, self-initiating resins are suitable for formulation of inks and varnishes for indirect food contact applications.

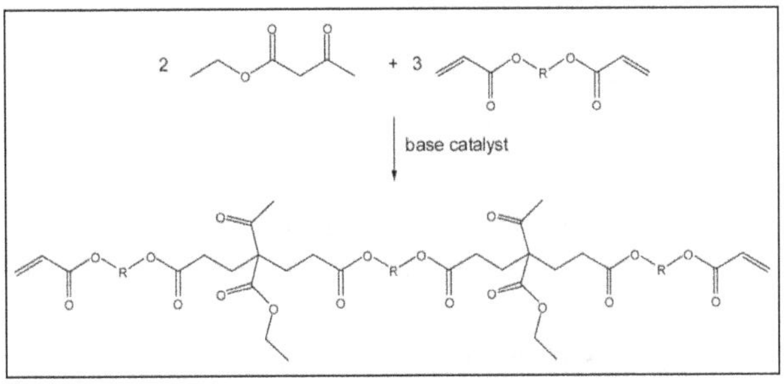

Figure 73: Self-initiating resins are obtained from Michael addition reaction between β-keto esters and diacrylates

3.3.2.9 Thiol-ene systems

Radiation curable thiol-ene systems occupy a special position among radically curable binders. Their crosslinking is based on the reaction between multifunctional olefins or allyl ethers ("enes") and thiols (radical polyaddition), and can be initiated by UV or electron-beam radiation. Figure 75 shows the course of the crosslinking reaction: under irradiation, a photoinitiator converts the thiol to a thiyl radical (step 1), which adds to the olefin molecule (step 2). The thioether radical formed can further react with an additional thiol via chain transfer (termination of the chain with formation of a new thiyl radical, step 3). In the presence of oxygen, hydroperoxides are formed (step 4). Because another thiyl radical originates in the process, the crosslinking reaction continues, and, in contrast to the polymerization of acrylates, is not inhibited by oxygen. In the absence of oxygen, the thioether radicals can recombine (step 5) [72, 73].

As Figure 75 shows, the reaction is based on a stepwise addition reaction. Chain growth occurs via chain transfer. For polymer formation, therefore, it is essential that the raw materials used are at least difunctional, and then produce linear chains. A network is obtained only with higher functionalities.

Commercially available ene components include, e.g., (cyclo)aliphatic olefins, vinyl ether, propenyl ether, allyl ether, acrylates, and lower-molecular polybutadiene oils. The thiol crosslinkers include aliphatic and aromatic dithiols, but also products derived from the reaction of polyols with mercaptopropionic acid, such as trimethylolpropane tris(mercaptopropionate)

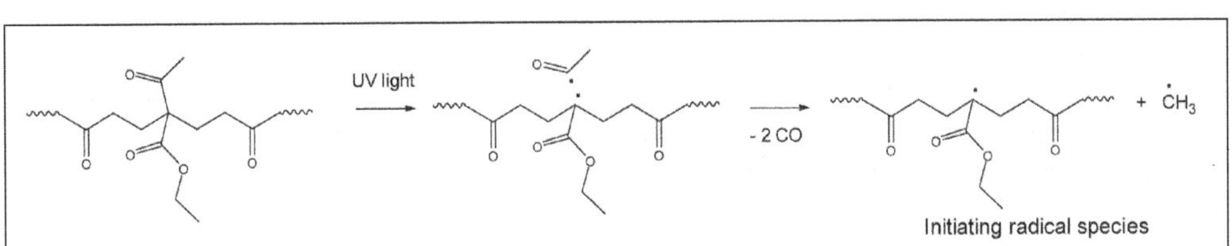

Figure 74: Proposed Norrish Type I cleavage of self initiating oligomers

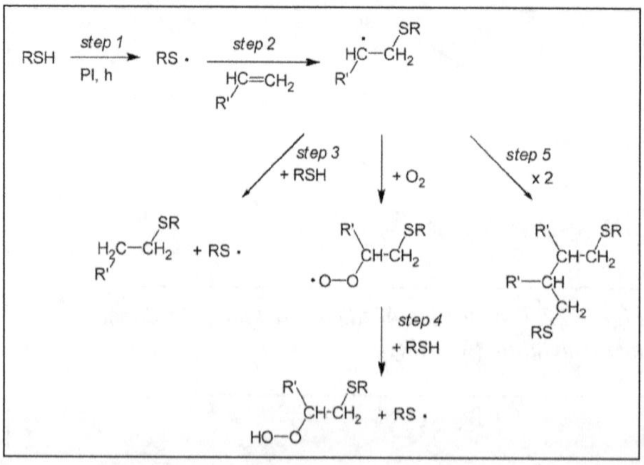

Figure 75: Reaction mechanism of thiol-ene polymerization; PI = photoinitiator

(F = 3) and pentaerythritol tetra(mercaptopropionate) (F = 4). The thiol-ene systems have a low viscosity and high crosslinking speed, even in the presence of oxygen. They have a low level of toxicity, and cause only mild skin irritation. The cured films have good heat resistance, flexibility, and adhesion especially to copper. The strong odor, particularly of the uncured formulations, is a disadvantage, as is the high price.

Areas of application

Thiol-ene systems are used in certain specific areas, e.g. for coating of printed circuit boards and electrical wires, in optical fibers, and for flexo printing plates.

3.3.2.10 Unsaturated polyesters

Chemical structure and preparation

Unsaturated polyesters (UP) are prepared by esterification and polycondensation of multifunctional alcohols and carboxylic acids. The polymerizable double bond is incorporated into the polymer by the use of unsaturated carboxylic acids such as maleic or fumaric acid (Figure 51). The properties are controlled via the molecular weight, the number of double bonds and the choice of monomers. In contrast to the acrylate functional oligomers, the polymerizable double bonds here lie along the polymer backbone. Due to this steric hindrance, reactivity is much lower compared to acrylates (cf. Figure 57). The linear or slightly branched resins are usually highly viscous and therefore solved in monomers; styrene is most often used, as it copolymerizes well with unsaturated polyesters during the curing process. Acrylic acid esters like TMPTA and vinyl ethers are also commonly used.

Properties

Unsaturated polyesters are used on account of their low price, good finish quality, and excellent filling power. The films are hard and resistant to solvents. Main drawbacks include low reactivity, strong inhibition by (atmospheric) oxygen, a tendency toward yellowing over the time, and the relatively poor mechanical properties of the film. The volatility of styrene, its odor, and occupational hygiene problems restrict their use. Oxygen inhibition during crosslinking can be reduced by modification of the resins with allyl ethers such as trimethylolpropane diallyl ether (high-gloss polyester) or by the use of paraffins, which float up to the surface, forming a matt protective layer.

Areas of application

Unsaturated polyesters are among the oldest radically curable resins known, and their continued use today is due almost entirely to their low price and easy availability. Because of their low reactivity, they are suitable only for UV curing, and not for EB curing, where the radiation dose required for satisfying cure would be too high. Almost without exception, they are used for wood and furniture coatings (mainly in southern and eastern Europe). Here they are used for leveling and to give body, for example in base coats, which are then covered with topcoats based on other oligomers. Because of the disadvantages mentioned above, unsaturated polyesters are increasingly being replaced by acrylate functional oligomers.

3.3.2.11 Saturated resins

Resins containing no polymerizable double bonds are also added to formulations to modify the mechanical properties and reduce costs. Examples are (chlorinated) polyesters, poly(meth) acrylates, and ketone-aldehyde resins. In general, the crosslink density is reduced, so that flexibility and adhesion are improved by the reduction of volume shrinkage. However, the reactivity and chemical resistance of the overall system are reduced, and high molecular varieties increase the viscosity. It appears to be possible to offset these disadvantages by using customized oligomers [37, 74]. Some of these "passive" resins are used to improve pigment wetting and stabilization.

In addition to resins that do not actually participate in the crosslinking reaction, a number of products exist, which, although saturated, can participate in crosslinking by abstracting hydrogen atoms (e.g., chlorinated rubber and chlorinated polypropylene).

3.3.3 Monomers for curing by free-radical polymerization

Monomers are classified according to their functionality as mono, di-, or polyfunctional. The viscosity of these low-molecular film-forming components increases with molecular weight and functionality, so that monofunctional monomers are good diluents, and polyfunctional monomers are used as low-viscosity components for tuning of special effects. Table 13 is intended as an aid to monomer selection.

Due to their low molecular weight compared to oligomers, a higher risk exists of sensitization and irritation of the skin, eyes, or respiratory tract. Direct contact must be strictly avoided (see Chapter 7). Additionally, the odor increases with increasing volatility.

Monomers allow enhancement of curing speed and, through variation in the crosslink density, control of such properties as hardness, flexibility, and chemical resistance of the film. A number of mono- and difunctional monomers cause surface swelling of various plastics such as polycarbonate, polystyrene, and ABS. The wetting of the substrate can be optimized via the surface tension of the monomers. Good wetting and surface swelling behavior form the basis of good adhesion to plastics (see Chapter 3.3.1.4.1).

Monomers that, apart from the polymerizable group, consist only of hydrocarbons are characterized by low surface tension and high resistance to water, yellowing, and weathering. They are usually prepared by reaction of the corresponding alcohols with acrylic acid, as shown in Figure 55. Some monomers are also available as methacrylates, which can be used to improve adhesion and toughness but have lower curing speeds [75]. Therefore, methacrylates are often used if a good adhesion to metals and plastics and excellent mechanical properties are important, e.g., industrial coatings.

Table 13: Influence of functionality of monomers on coatings properties [76]

<table>
<tr><th>Functionality</th><th>Reactivity</th><th>Shrinkage</th><th>Adhesion</th><th>Flexibility</th><th>Tensile strength</th><th>Solvent resistance</th></tr>
<tr><td>mono (1)</td><td rowspan="4">▲</td><td rowspan="4">▲</td><td rowspan="4">▼</td><td rowspan="4">▼</td><td rowspan="4">▲</td><td rowspan="4">▲</td></tr>
<tr><td>di (2)</td></tr>
<tr><td>tri (3)</td></tr>
<tr><td>tetra (4)</td></tr>
</table>

Table 14: Properties of alkoxylated homologues of HDDA compared with HDDA [22]

Name	Surface tension/mN/m	Volume shrinkage/%	Draize value	Pendulum hardness of the cured film
hexanediol diacrylate (HDDA)	36.0	12.3 13.1 [44]	6.0	58
ethoxylated hexanediol diacrylate ($HD(EO)_{2.0}DA$)	37.8	9.5	1.7	50
ethoxylated hexanediol diacrylate ($HD(EO)_{3.0}DA$)	38.4	7.3	–	33
ethoxylated hexanediol diacrylate ($HD(EO)_{3.7}DA$)	39.1	7.1	–	31
propoxylated hexanediol diacrylate ($HD(PO)_{2.0}DA$)	38.8	6.3	1.2	18

To reduce volatility and the associated odor and health risks, higher-molecular monomers have been developed by alkoxylation of the corresponding alcohols with ethylene oxide (EO) or propylene oxide (PO) and subsequent acrylation. A wide variety of these components are now commercially available, most of them carrying short-chain oligoether units (2 to 15 EO or PO units). The transition from monomer to oligomer is not exactly defined here (cf. polyether acrylates see Chapter 3.3.2.4). Despite their higher molecular weight, which reduces volume shrinkage during the curing process, alkoxylated monomers reduce viscosities of formulations very well. The higher polarity of ethoxylated monomers improves their dissolving power for urethane, polyester, and polyether acrylates, but reduces water resistance. The presence of ether groups leads to greatly improved (surface) curing and flexibility, but to a reduced resistance to yellowing and weathering (see Chapter 3.3.2.4). Propoxylated monomers have good mechanical properties and good waterresistance. Some of them have low surface tension, and can therefore be used to improve adhesion on nonpolar substrates (e.g., propoxylated neopentyl glyol diacrylate ($NPG(PO)_2DA$), 31.5 mN/m). Table 14 shows some properties of alkoxylated homologues of HDDA [77]. Vinyl ethers are also used in radically curable systems, and are described in Chapter 3.3.4.2. Some commonly used monomers are described below.

Note

Especially in the case of di- or polyfunctional monomers, commercially available products deviate from the idealized chemical structures due to side reactions during their production (mainly by *Michael* addition). Commercially available products are therefore usually not single components but mixtures of various components. Figure 76 shows the formation of the TMPTA-TMPDA adduct from trimethylolpropane triacrylate (TMPTA) and trimethylolpropane diacrylate (TMPDA), both of which can be found, among other components, in commercially available products.

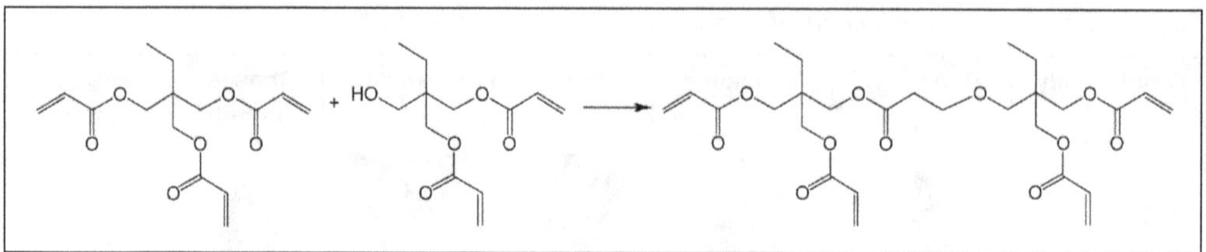

Figure 76: Side reactions occur during the preparation of the monomers: formation of TMPTA-TMPDA adduct

Table 15: Properties of selected monofunctional acrylate monomers

Name (Acronym)	Viscosity /mPa · s @ 25 °C	Properties/areas of application
cyclic trimethylolpropane formal monoacrylate (CTFA)	10 to 20	slight odor, low volume shrinkage, resistant to yellowing, good chemical resistance and hardness; used in coatings for wood, plastics, and metals, and in printing inks
2-(2-ethoxyethoxy) ethyl acrylate (EOEOEA)	5 to 10	high reactivity, good dissolving power, low Tg uses: plastic coatings, PVC flooring
isobornyl acrylate (IBOA)	5 to 15	high T_g, good weathering and abrasion resistance. Problem: odor uses: coating of plastics, wood, and metals; OPVs, screen printing, and the electronics industry
isodecyl acrylate (IDA)	5 to 10	low surface tension and Tg, good wetting of plastic substrates, good water resistance uses: coatings on polyolefins, PVC flooring, printing inks
octyl/decyl acrylate (ODA)	5 to 10	good wetting of plastics, high water resistance uses: coatings and screen inks
2-(phenoxy)ethyl acrylate (2-PEA)	5 to 15	non-irritant, pungent odor, good solvent properties, high reactivity uses: coatings for plastics and metals, screen printing
tetrahydrofurfuryl acrylate (THFA)	5 to 10	good (polar) solvent properties, high reactivity uses: plastic coatings, PVC flooring/sheeting

3.3.3.1 *Monofunctional monomers*

Monofunctional monomers are used in quantities of less than 15 % relative to the total formulation to reduce viscosity. By reducing the crosslink density, they improve flexibility and adhesion by reducing volume shrinkage on non-absorbent substrates (e.g., metals and plastics such as polyolefins and PMMA) and swellable substrates (plastics such as ABS, PVC, and PC). However, the low crosslink density results in poorer resistance to chemicals and mechanical stress. Table 15 provides an overview of some further properties of selected monofunctional monomers.

3.3.3.2 *Difunctional monomers*

Difunctional monomers are the most important reactive diluents. They have good reactivity and a good hardness-elasticity balance. They are used in amounts of between 15 and 60 % of the total formulation.

3.3.3.3 *Polyfunctional monomers*

Most polyfunctional monomers have functionality between three and six. They are essentially used in greater or lesser quantities for tuning certain properties, e.g., to increase curing speed, rather than for reducing viscosities. Their high functionality results in a high crosslink density that improves hardness and scratch and chemical resistance. Table 17 shows other properties of selected members of this monomer class.

Table 16: Properties of selected difunctional acrylate monomers

Name/acronym	Viscosity /mPa · s @ 25 °C	Properties/areas of application
dipropylene glycol diacrylate (DPGDA)	5 to 15	good dissolving power, good adhesion very widely used in coatings and printing inks
ethoxylated bisphenol A diacrylate (BPA(EO)$_4$DA)	900 to 950	low-viscosity "oligomer", good hardness and scratch resistance uses: wood and plastic coatings for spray application
ethoxylated hexanediol diacrylate (HD(EO)$_3$DA)	25 to 35	good flexibility, low volume shrinkage used as alternative to HDDA in coatings for wood, metals, and plastics, and in flexo printing inks
hexanediol diacrylate (HDDA)	5 to 10	good dissolving power, good adhesion to PVC, PC, and ABS, good flexibility, good weathering stability uses: PVC flooring, wood coatings, OPVs, printing inks
propoxylated neopentyl glycol diacrylate (NPG(PO)$_2$DA)	10 to 20	low shrinkage, good pigment wetting, high flexibility, good substrate wetting (low surface tension) uses: plastic coatings, wood topcoats, flexo and inkjet inks, release coatings
tripropylene glycol diacrylate (TPGDA)	5 to 15	very commonly used in coatings and printing inks

Table 17: Properties of selected polyfunctional acrylate monomers

Name/acronym	Viscosity /mPa · s @ 25 °C	Properties/areas of application
dipentaerythritol pentaacrylate (DiPEPA)	7000 to 14,000	uses: printing inks
di-trimethylolpropane tetraacrylate (DiTMPTTA)	500 to 700	good adhesion uses: litho and screen inks
ethoxylated trimethylolpropane triacrylate (TMP(EO)$_3$TA)	40 to 80	good hardness-elasticity balance, good adhesion uses: coatings, flexo and screen inks
pentaerythritol tetraacrylate (PETTA) (often supplied in a form containing pentaerythritol triacrylate)	400 to waxy	good adhesion to plastics uses: coatings for metals, wood, and plastics; printing inks (flexo, gravure, offset, screen)
propoxylated glycerol triacrylate (G(PO)$_3$TA)	70 to 100	good hardness-elasticity balance, good pigment wetting widely used in wood coatings, OPVs, and printing inks
trimethylolpropane triacrylate (TMPTA)	70 to 110	good resistance to heat and weathering, high gloss very widely used in coatings and printing inks
tris(2-hydroxyethyl) isocyanurate triacrylate (THEICTA)	solid, m.p. 53 °C	high T_g, very good heat resistance, good pigment wetting and adhesion used in flexo printing inks and topcoats for wood and plastics.

3.3.3.4 *Inert solvents*

Small quantities of inert organic solvents are used in some formulations to improve flow behavior and deaeration. Additionally, the reaction rate and degree of cure may be increased due to the increased mobility of the film-forming components. The softening action of the solvents reduces the glass transition temperature of the film being formed. Internal stresses can thus be reduced, improving adhesion (see Chapter 3.3.1.4). Some high-boiling solvents have low vapor pressure and do not contribute to the VOC content.

3.3.4 *Resins for curing by cationic polymerization*

Sales volumes of cationically curable formulations have been more than doubled between 2002 and 2005. The key growth areas are printing inks, packaging, and coatings for plastics and metals [78]. Cycloaliphatic epoxides are generally used as oligomers, while properties of formulations and resulting films are modified by vinyl ethers and oxetanes.

3.3.4.1 *Epoxides*

Cycloaliphatic epoxides are used in cationically curable formulations in quantities of up to 70 %. Most systems contain the classic ECC (3,4-epoxycyclohexylmethyl 3',4'-epoxycyclohexane carboxylate), which is reactive and has a low viscosity (Figure 77). The cured films exhibit high toughness and resistance to solvents, and adhere well to a wide range of substrates. The high brittleness is a disadvantage, however. ECC is used in coatings for metals and plastics, and also for overprint varnishes and flexo printing inks.

All other cycloaliphatic epoxides are used mainly to adapt the properties of formulations containing ECC to the respective application. For example, flexibility can be improved by varying the spacers between the epoxycyclohexyl rings. Examples include bis(3,4-epoxycyclohexylmethyl) adipate and transesterification products of methyl-3,4-epoxycyclohexane carboxylate and polyols (Figure 78). The possibilities for variation are wide due to the large number of linear or branched (cyclo)aliphatic polyols that can be used [79]. Areas of application include metal and plastic coatings, overprint varnishes, and flexible packaging.

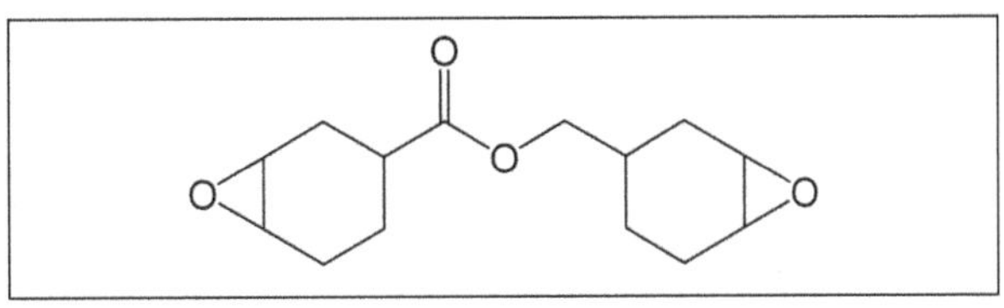

Figure 77: ECC (3,4-epoxycyclohexylmethyl 3',4'-epoxycyclohexane carboxylate) is widely used in cationically curable formulations

In addition to cycloaliphatic epoxides, other epoxides are also used, essentially to reduce costs and enhance flexibility. Examples include epoxidized oils, such as soy oils, and epoxidized polybutadiene oils, which also increase chemical resistance and toughness. Costs are also reduced by the use of conventional bisphenol A diglycidyl ethers, which have lower reactivity, and epoxidized novolacs [34, 36]. The viscosity can be reduced by using, for example,

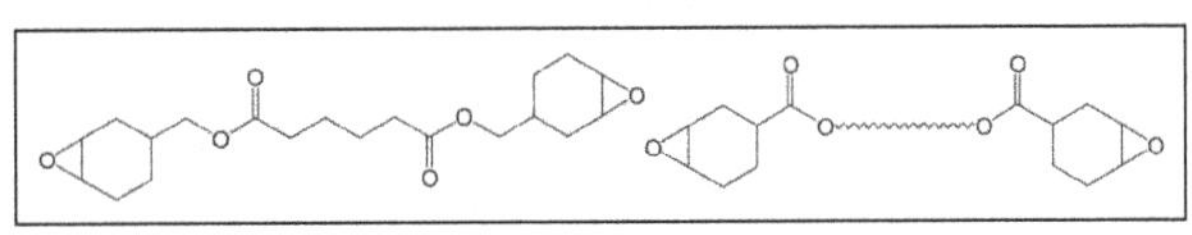

Figure 78: Left: bis(3,4-epoxycyclohexylmethyl) adipate; right: transesterification products of methyl-3,4-epoxycyclohexane carboxylate and linear or branched (cyclo)aliphatic polyols

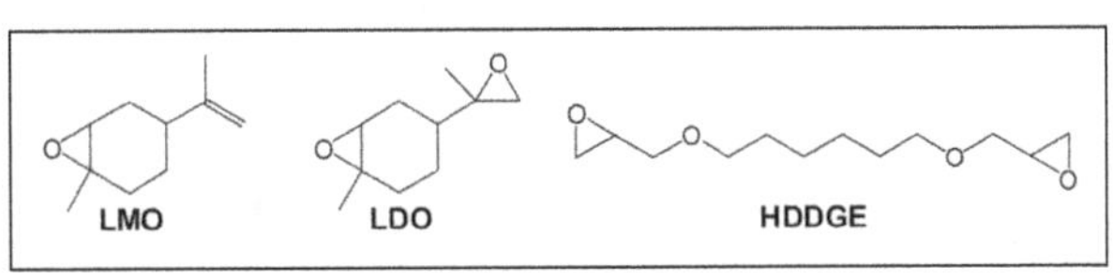

Figure 79: Limonene monoxide (LMO), limonene dioxide (LDO), and 1,6-hexanediol diglycidyl ether (HDDGE)

1,6-hexanediol diglycidyl ether (HDDGE) or by limonene monoxide or dioxide (LMO, LDO), which have good reactivity and dissolving power (Figure 79).

3.3.4.2 *Vinyl ethers*

In combination with epoxides, vinyl ethers are used in quantities of about 10 to 15 % (and in some cases up to 25 %) to reduce viscosity in printing inks and overprint varnishes. In these applications they increase reactivity and therefore curing speed. They most likely do not copolymerize with epoxides, but form interpenetrating networks [80]. In general, the vinyl ether groups do not react completely during the cure, and remaining groups form aldehydes as a result of hydrolysis. Yellowing resistance is therefore low.

The usually non-irritant compounds are prepared, for example, by the *Reppe* synthesis, involving the addition of acetylene to alcohols. Typical representatives of the group are the low-viscosity di- and triethylene glycol divinyl ethers, polytetrahydrofuran divinyl ether, 4-hydroxybutyl vinyl ether, n-dodecyl vinyl ether, and 2-ethyl-1-hexyl vinyl ether, which are used to increase curing speed, flexibility, and adhesion. Monomers, such as cyclohexanedimethanol divinyl ether, increase hardness and solvent resistance, but at the expense of flexibility.

> Note
>
> Due to their high diluting power, vinyl ethers like triethylene glycol divinyl ether can also be combined in quantities of up to 10 % with radically curable acrylates and in quantities of up to 40 % with unsaturated polyesters to reduce viscosity. In this case, they reduce inhibition by oxygen, improving the scratch resistance. Labeling is not necessary due to their lower toxicity. Their copolymerization parameters are such that they cannot be radically homopolymerized, and require the above-mentioned reactants.

3.3.4.3 *Oxetanes*

Like epoxides, oxetanes lead to low volume shrinkage because they also cure via a ring-opening polymerization. For example, 3,3-dimethyl-2-p-methoxyphenyloxetane (MPO) and trimethylolpropane oxetane (TMPO) are used as non-irritant diluents with good dissolving power for coatings and printing inks (Figure 80) [36]. The films have good solvent resistance and adhesion. TMPO flexibilizes epoxy films and accelerates cure due to the presence of a hydroxyl group in addition to the oxetane ring (see Chapter 3.3.4.4).

3.3.4.4 *Polyols*

Hydroxy functional components such as polycaprolactone, polyether, polyester, and polycarbonate polyols are used in quantities of up to 30 % on the total formulation for flexibilization and improvement of toughness. They crosslink by chain transfer [36, 81]. It has been found in practice that the ratio of epoxy to hydroxyl groups should lie between 1.5 and 10 to obtain films with good mechanical resistance [79, 80]. A high ratio improves surface cure and therefore hardness, as well as solvent resistance and adhesion, but has a negative effect on through-cure, flexibility, and impact resistance.

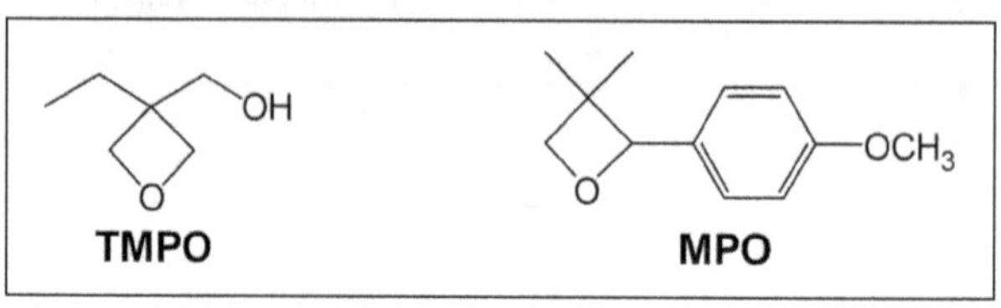

Figure 80: Trimethylolpropane oxetane (TMPO) and 3,3-Dimethyl-2-p-methoxyphenyloxetane (MPO)

3.3.4.5 *Acrylates*

Most of the currently used photoinitiators for cationic polymerization produce free radicals that can initiate polymerization of acrylate functional compounds. In the simultaneous presence of cationically and radically curable compounds, interpenetrating networks are formed from the variously crosslinked polymers, allowing the properties of both systems to be combined (hybrid systems). Free-radical and cationic photoinitiators are usually combined to speed up curing [82]. Because cationic polymerization is slower than radical, however, complete conversion of all the components of the hybrid system is usually difficult to achieve. Cationic curable formulations containing small amounts of acrylates have now become developed.

3.4 Special radiation curable systems

Low-molecular monomers used in liquid RC coatings and printing inks have distinct disadvantages like volatility, odor, and toxiticity. These limitations have driven the development of alternative systems like radiation curable water-borne and powder coatings. Due to the high molecular weight of the oligomers used in these systems, the volatility is lower than in 100 % liquid RC systems. Moreover, the use of monomers is not necessary, so that considerations of extractability of unreacted monomers and adhesion problems resulting from high volume shrinkage do not arise. Radiation curable water-borne and powder coatings combine the flexibility of the higher-molecular oligomers with the hardness and resistance resulting from the high crosslink density through the acrylate groups [83, 84]. Radiation curable water-based, powder, and additionally dual-cure formulations are treated under "Special Systems". While the water-based RC coatings are already well established, the powder and dual cure technology are still niche areas.

3.4.1 Water-based radiation curable coatings

While conventional 100 % liquid RC systems suffer from the drawbacks described above, properties such as gloss, scratch and chemical resistance of many water-based coatings are in many cases inadequate. Thermosetting, aqueous systems that crosslink chemically also have long cure times at elevated temperatures. A merging of the two technologies combines their advantages while reducing the disadvantages. The main difference from 100% liquid RC coatings lies in the avoidance of monomers as reactive diluents, and in the use of high molecular weight binders, usually present in the form of aqueous emulsions or dispersions, which lead to good mechanical properties and adhesion.

After application of the aqueous formulation, the water has to be removed first. This uncured film may already be capable of withstanding stress. However, the final properties are obtained only after photocrosslinking. Table 18 summarizes the advantages and disadvantages of water-based RC systems compared with conventional 100 % liquid RC formulations and conventional aqueous dispersions [85, 86, 87].

Aqueous polyester, polyether, polyacrylate, and polyurethane acrylates are usually used as binders. In practice, a distinction is made between emulsions and dispersions, although the terminology as used here is, strictly speaking, inaccurate. In physics, a dispersion is the generic term for two phases, one of which is distributed, but not dissolved, in the other. An emulsion is a mixture of two liquids that are not soluble in each other (e.g., an insoluble oligomer in water). The coatings industry uses these terms to describe the film obtained after evaporation of water. Emulsions result in tacky films, whereas dispersions produce tack-free films prior to UV curing [86]. This means that resins used in emulsions have a low T_g and comparatively low molecular weights, while those in dispersions have high (M_w up to 50,000 g/mol) to very high (M_w greater than 100,000 g/mol) molecular weights and higher T_g.

The VOC content of emulsions is low because it is not necessary to use auxiliary solvents for film formation and good flow. Furthermore, recycling and cleaning of working equipment

Table 18: Advantages and disadvantages of water-based RC coatings

Advantages
• Low viscosity, independent on the molecular weight of the binders (for dispersions) • Good adhesion (reduced volume shrinkage) on metals, plastics, and particularly wood • Very good hardness-elasticity balance due to high molecular weight (better scratch and chemical resistance than conventional water-based systems) • No monomer necessary/zero solvent emission • Reduced risk of sensitization and of irritation of respiratory tract, eyes and skin • Low odor • Low tendency to migration and exudation of monomers • Reduced flammability and explosion risk • Better mattability than 100 % liquid UV systems and higher gloss than conventional water-based systems • Existing equipment used for application (e.g., spraying, curtain, rolling) • After removal of the water, the films usually withstand stress well and can be repaired before curing • In case of wood coatings: excellent substrate wetting, excellent reproduction of the substrate structure • Equipment can be cleaned with water
Disadvantages
• Removal of water prior to UV curing needs energy, time (curing rates lie between those for 100 % liquid RC and thermosetting aqueous systems), and space • Use of amines increases VOC content • Application and curing must be carried out under defined climatic conditions (temperature, low atmospheric humidity) • Gloss lower than for 100 % liquid UV systems

are easily carried out. The disadvantages include the slow evaporation of water, the low stress resistance of the water-free films before crosslinking, and poor mattability. Adhesion and mechanical resistance are sometimes only moderate, and high concentrations of photoinitiators are often necessary to crosslink the material sufficiently. Typically, emulsions of allyl polyester, polyester acrylates, polyether acrylates, and polyurethane acrylates are used in the coating industry [88].

By contrast, dispersions exhibit a significantly better hardness-elasticity balance, improved adhesion and chemical resistance, better substrate wetting, and good mattability. Moreover, lower concentrations of photoinitators are required. The water also evaporates faster and the uncured water-free films are already capable of withstanding stress. Main limitation is the flow, which often has to be improved by means of additives or (reactive) co-solvents. Three types of dispersions are commonly used [88]:

- blends of saturated polyurethane and/or polyacrylate dispersions with monomers
- dispersions of medium molecular weight polyurethane acrylates, and
- dispersions of high molecular weight polyurethane acrylates.

Figure 81 shows the relative properties of emulsions and dispersions [88].

Preparation

Some of the oligomers already described, such as many of the polyether acrylates and monomers, are water soluble or can be stably transferred into the aqueous phase by the addition of emulsifiers (external hydrophilization). In the case of emulsions, for example, a polyester acrylate with low T_g and free carboxylic acid groups is neutralized by a base (e.g., an amine) and converted from the melt to an aqueous emulsion by addition of water. To introduce the

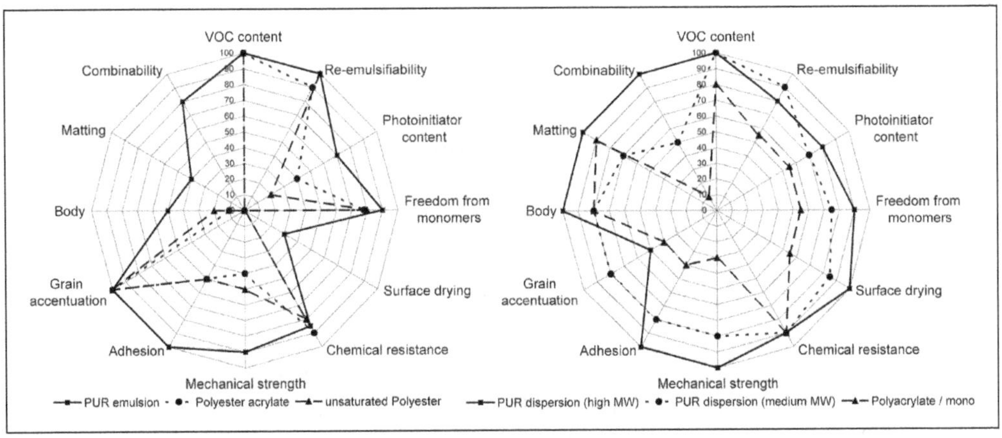

Figure 81: Relative properties of emulsions (left) and dispersions (right) [88]

acid groups, tricarboxylic acids such as trimellitic anhydride can be used in polyester synthesis when only two of the three acid functions are esterified [89].

Saturated polyurethanes are prepared as shown in Figure 82. For ionic hydrophilization dimethylol propionic acid (DMPA) is often incorporated, and is neutralized by an amine. Because the neutralizing agent leaves the film during the evaporation of the water, high water resistances are obtained here [89]. Internal hydrophilization can also be achieved by incorporating polyethers. In this case the films may have permanent water sensitivity but are free from neutralizing agents (zero VOC content). The saturated resins usually have

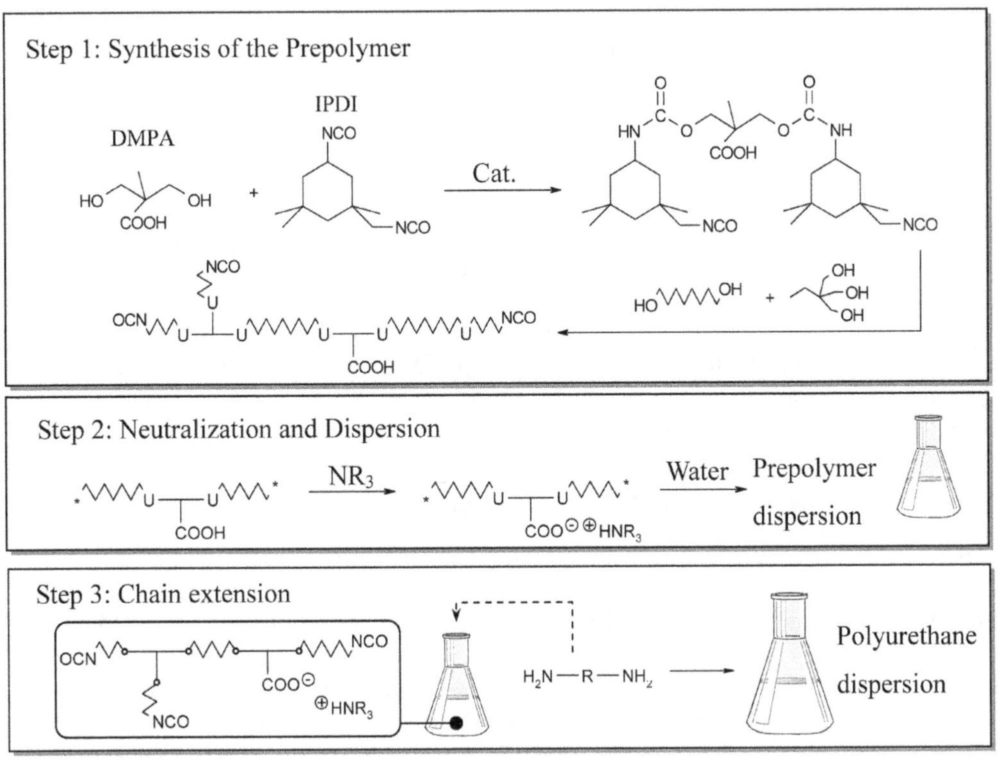

Figure 82: Preparation of saturated polyurethane dispersions based on isophorone diisocyanate (IPDI) and dimethylol propionic acid (DMPA); U = urethane group, HO〰OH = sarated polymeric diol [89]

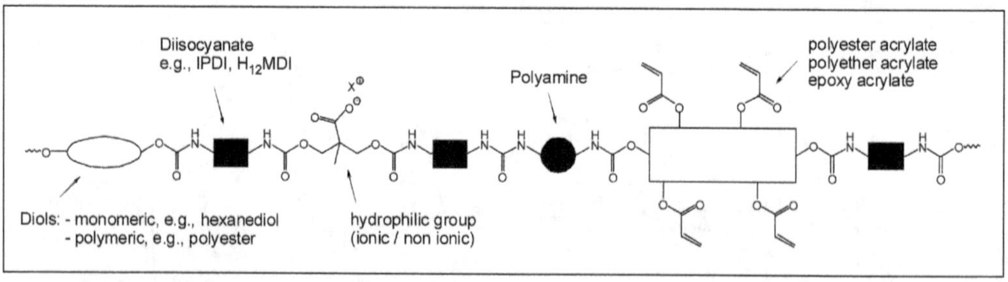

Figure 83: Unsaturated polyurethane dispersions are based on polymeric acrylates [88]

high molecular weights. For a minimum of UV crosslinking reactive diluents (monomers) are added resulting in low crosslink densities after UV cure. Mechanical and resistance properties are therefore often only moderately good.

Due to the high (melt) viscosities resulting from their high molecular weights, PU prepolymers often cannot be prepared from the melt; co-solvents are required, which either remain in the dispersions (like reactive diluents or (toxic) N-methyl pyrrolidone) or are removed, after the neutralized resin has been transferred into the aqueous phase, to give VOC-free dispersions ("acetone method") [85]. Because the latter process is energy- and time-intensive, PU dispersions prepared in this way are more expensive.

The synthesis of medium molecular weight polyurethane acrylates is similar to that of the saturated polyurethanes but monomeric hydroxyalkyl acrylates are incorporated to make the resins suitable for UV crosslinking. Due to the low density of acrylic double bonds reactive diluents are also used here to improve mechanical and resistance properties [88]. Because of the negative impact of reactive diluents on the storage stability of dispersions polyurethane acrylate dispersions of high molecular weight have been developed what also addresses odor and hygiene issues. They contain high acrylate functionalities resulting in high crosslink densities without additional reactive diluents. Their synthesis is similar to that described above, but acrylate functional polymers like polyester, polyether, or epoxy acrylates are incorporated (Figure 83).

Areas of application

Water-based radiation curable coatings are being increasingly used as primers or topcoats for wood, e.g., for furniture, kitchen units, and wooden window frames and doors. They have excellent adhesion to wood resulting from excellent wetting and from alignment of the wood fibers increasing the contact surface area [87]. Additionally, the resins and oils contained in wood are not attacked by the low-molecular species that are present in 100 % liquid RC systems. Water-based RC systems are also used in coatings for metal and plastics, and in some printing ink applications [86, 90].

3.4.2 Radiation curable powder coatings

Conventional powder coatings are simultaneously melted and cured at temperatures between 130 and 240 °C. This process can take up to 30 minutes. Particularly for low-temperature curing powder coatings, crosslinking occurs during the melting process, which often results in flow that is less than optimal (orange peel effect). The high curing temperature restricts the area of application of powder coatings to heat-resistant substrates like glass or metal. The new technology of UV curable powder coatings combines the advantages of UV curing with those of powder coatings. Because the melting process is carried out separately from the curing, extremely smooth surfaces are obtained. Melting and leveling are effected by IR radiation below 100 °C and take only a few minutes. Crosslinking occurs within a

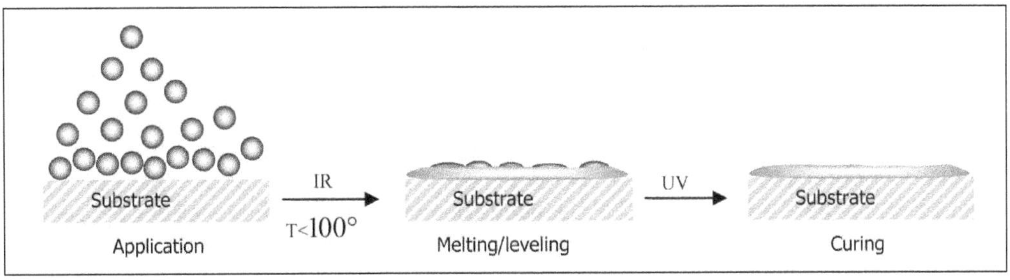

Figure 84: Melting and curing of UV powder coatings are carried out in separate steps

few seconds by UV radiation (Figure 84). The low melting temperature allows even heat-sensitive substrates like wood or plastics to be coated.

UV powder coatings must have low melting temperatures and, to ensure good flow of the melt, low viscosity. However, storage stability up to 40 °C is also necessary. Because these parameters depend on the glass transition temperature of the formulation, this opens up a gap that can be closed by combining amorphous and (partly) crystalline polyester urethane acrylates [84]. In addition to urethane acrylates, acrylated polyacrylates and polyester (meth) acrylates are also used as binders.

UV powder coatings are currently used commercially in only a few applications. Examples include coating of temperature-sensitive substrates for MDF furniture and PVC flooring, where UV powder coatings provide very good mechanical and optical properties. Thanks to these advantages, many other applications will undoubtedly follow. Table 19 shows the advantages and disadvantages of this technology as compared with conventional radiation-curable coatings and thermosetting powder coatings.

3.4.3 Dual cure technology

The technologies described above are restricted to applications on elements having simple shape, allowing UV light to reach the whole surface of the substrate. For this reason, RC systems are used mainly to coat flat objects. In many applications, however, complex,

Table 19: Advantages and disadvantages of UV powder coatings [84]

Advantages
• Avoidance of irritant monomers (reduced risk of sensitization and irritation of respiratory tract, eyes, and skin; low odor, no VOC) • Good adhesion (reduced volume shrinkage) • Good flow/smooth surface structure • Suitable for temperature-sensitive substrates (plastics, wood, MDF, paper, etc.) • Good hardness-elasticity balance and chemical resistance • Relatively low film thickness (as compared with thermosetting powder coatings) • Low space requirements during application and cure • Faster curing speed (as compared with thermosetting powder coatings) • Simple recycling (of overspray)
Disadvantages
• Strict adherence to application and cure conditions is necessary • Flow not yet optimal on some types of wood (in the region of the edge of the workpiece) • Storage stability • Poor mattability

three-dimensional coated objects have shadow areas light hardly reaches, e.g., automotive bodywork (see Chapter 5.4.1). One solution is to use special curing equipment like plasma technology (see Chapter 4.1.2.4). Alternatively, a minimum of crosslink density in the shadow regions can be achieved by the use of UV curable coatings that simultaneously cure by a radiation-independent mechanism (dual cure, DC) [91]. Such DC systems can also be used for curing of pigmented formulations when the absorption range of the pigment overlaps with that of the photoinitiator limiting the penetration depth of light into a thick pigmented section. Additional reasons for using RC coatings that crosslink by a further radiation-independent mechanism are as follows [91]:

- improvement of adhesion on non-absorbent or non-swelling substrates by reduction of internal stresses and volume shrinkage,
- improvement of adhesion on, e.g., oily woods by reaction with the substrate,
- increase of elasticity, and
- production of deformable, elastomeric sheets or coils, which attain their final strength by UV exposure after the forming step.

In principle, any curing mechanism for conventional coatings is suitable as the additional reaction step. So far, however, only the curing of polyisocyanates with polyols (1K or 2K PURs) has been of practical significance [91]. The following description will therefore concentrate on UV PUR dual cure.

The oligomers and monomers described previously can be used as film-forming components if they contain hydroxyl as well as acrylate groups. It is absolutely necessary for each molecule to have both functionalities so as to be able to crosslink both radically as well as with isocyanates. The absence of an OH group can impair resistance properties and even lead to tacky films in shadow regions. Customized oligomers can overcome these deficiencies. Examples include epoxy acrylates with secondary OH groups of low reactivity, and particularly urethane or polyester acrylates containing more reactive hydroxyl groups. If resins are used that simultaneously contain large numbers of acrylate and particularly isocyanate groups, typical coating polyols without acrylate groups can also be utilized in the formulation.

OH-functional, monomeric acrylates prepared from acrylic acid and polyfunctional alcohols such as trimethylolpropane or pentaerythritol can be used to reduce the viscosity. However, statistical esterification can lead to fully acrylated products, which can negatively impact resistance properties in shadow regions. For this reason, the viscosity is usually adjusted by adding up to 40 % of volatile organic solvents, which also support deaeration and flow (see Chapter 3.3.3.4) [92]. Special acrylate-group containing polyisocyanates have been developed for use in DC applications. These are, for example, reaction products of hydroxyalkyl acrylates with oligomeric or polymeric derivatives of diisocyanates. Various derivatives of hexamethylene diisocyanate are also suitable for this purpose [91].

Most DC systems are two-component (2K) coatings with a limited pot-life, due to the presence of hydroxyl and isocyanate groups. One-component (1K) systems with blocked isocyanates that deblock at moderately elevated temperatures (120 to 150 °C) have unlimited pot-life. In addition to these products, water-based formulations without VOC have been developed.

After application of the DC formulations, the two curing steps may be carried out in either order. A productivity advantage is obtained if the fast radiation curing is performed first, followed by the slower OH-NCO reaction (Figure 85). Regardless of the order chosen, it must be remembered that the mobility of the reactants for the second curing step is reduced in the partially cured film, which reduces the rate of curing in this second step (see Chapter 3.3.1).

A wider use of DC coatings is restricted by the high proportion of volatile organic solvents (VOC), the limited pot-life of the 2K systems, and the expensive raw materials. Because the

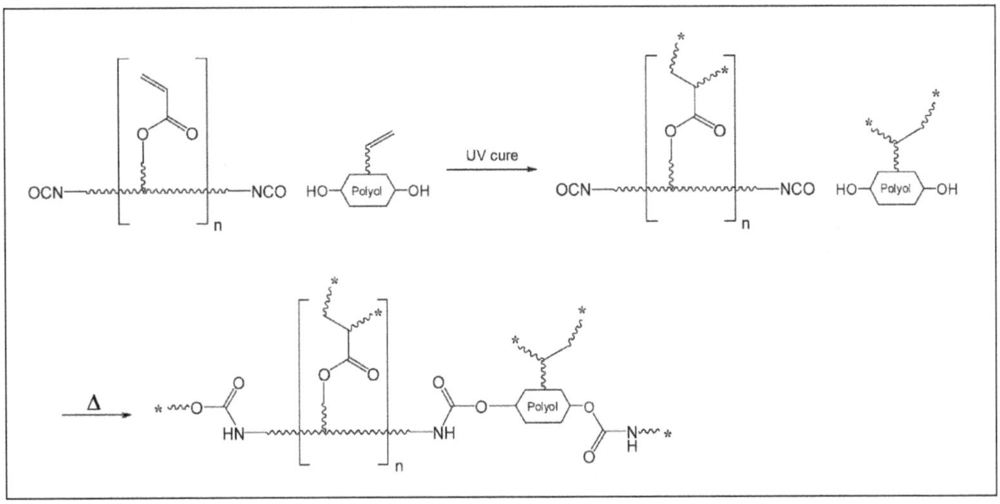

Figure 85: Curing process of a DC coating. The initial radiation curing is followed by the thermally induced reaction between the isocyanate and polyol components

solvent must be evaporated after application of the formulation and an additional curing step is required, productivity, while higher than for conventional coatings, is not as high as in purely radiation curable 100 % liquid coatings.

Some areas of application

- coating of wood substrates with high oil content (e.g., teak, mahogany)
- coating of porous materials such as cement or concrete, where excessively deep penetration of the coating into the substrate prevents complete through-cure
- coating of complex three-dimensional objects with shadow regions (e.g., OEM automotive bodywork with high scratch resistance)

3.5 Additives

Radiation curing coatings and inks display excellent properties in the final application. The formulation, processing and application however create challenges for the formulator. Unlike with other formulations one important influencing formulation component is not given in the case of a 100 % liquid RC formulation - the solvent.

The limited formulation freedom and the inherently high surface tension, relatively high viscosity and rather poor wetting properties of RC formulations make the use of additives truly important. Additives cannot solve every coating or ink problem, but they help to optimize the coating and ink properties.

The choice of additives however can be difficult as well, since the solvating power of solvent-free RC formulations is significantly lower than e.g. of solvent-based formulations. This may lead to poor additive solubility. Moreover, RC formulations are quite polar compared to solvent-based formulations and may display incompatibilities with non-polar additives. Consequently many traditional additives are unsuitable for RC coatings and inks. But with the continuously increasing importance of these formulations the additive suppliers have developed additives specifically for radiation curing. These products display a fine-tuned polarity profile resulting in good compatibility and performance. Often these products are supplied solvent-free to avoid any undesired VOC to the formulation. And following the route of only having crosslinkable components in a radiation curing formulation, more and more additives carry reactive groups (e.g., acrylate groups) which make them polymerizable.

Now the word additive is a very general term. It describes all kind of substances that are added to the basic formulation at relatively low addition levels (0.02 to 5 % on total formula weight) in order to improve certain properties. Mostly, additives are used to improve processing, storage, and application properties of the coatings and inks (properties before cure) and others that mainly influence the properties of the cured film (properties after cure).

Examples for additives affecting the properties before cure:
- pigment wetting and dispersing additives
- rheological additives
- substrate wetting additives
- defoamers and deaerators
- antioxidants/formulation stabilizers

Examples for additives affecting the properties after cure:
- surface control additives (flow, leveling, and slip, scratch resistance)
  - flow and leveling additives
  - waxes
  - silicone-based additives
- matting agents
- additives to improve adhesion
  - reactive adhesion promoters
  - thermoplastic co-binders

On the following pages these groups of additives will be briefly described, their function will be explained and mentioned what is special in radiation curing formulations. Guidance will be given for the selection of additives.

3.5.1 Pigment wetting and dispersing additives

In principle the production of a pigmented RC formulation is not much different to water-based or solvent-based formulations. Efficient pigment dispersion is of vital importance for the final coating or ink properties, such as color strength, gloss, and hiding power respectively transparency and last but not least rheology behavior. But it is not easy to achieve good pigment wetting and dispersion in solvent-free grinds based on monomers and oligomers. Wetting is slower and high pigment loadings are difficult to achieve since viscosity builds up easily with increasing pigment content. Especially in grinds of organic or carbon black pigments a strong thixotropy may already be observed below the desired pigment loading. The addition of suitable pigment wetting and dispersing additives before the dispersion process helps to improve dispersing a pigment by high mechanical energy entrained by high speed mixer, three roller mill, beads in a bead mill or other grinding media [93].

Before the dispersion process pigment powders mainly consist of agglomerates. The mechanical dispersion destroys the agglomerates and creates aggregates, to much lesser degree primary particles (Figure 86). Without sufficient stabilization these form very large particles again, the so called flocculates.

What the additives really do in the process of pigment dispersion is to support the pigment wetting before and during the grinding and to stabilize the achieved dispersion. Thus they prevent reflocculation of the dispersed pigment or filler particles. In literature two basic mechanisms of pigment stabilization are discussed [94] (Figure 87):
- electrostatic repulsion and
- steric hindrance.

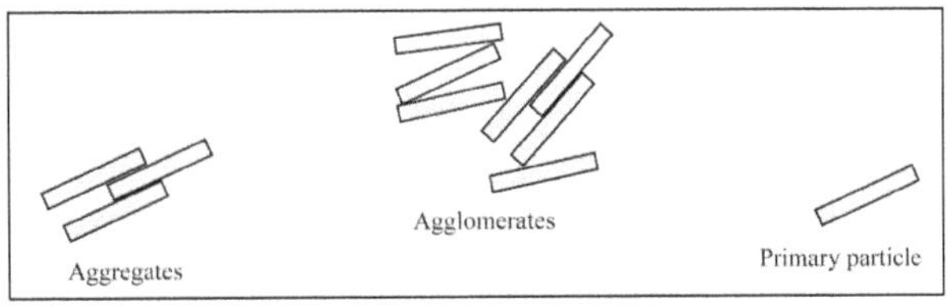

Figure 86: Possible aggregation forms of pigment particles in a coating formulation, source EVONIK TEGO

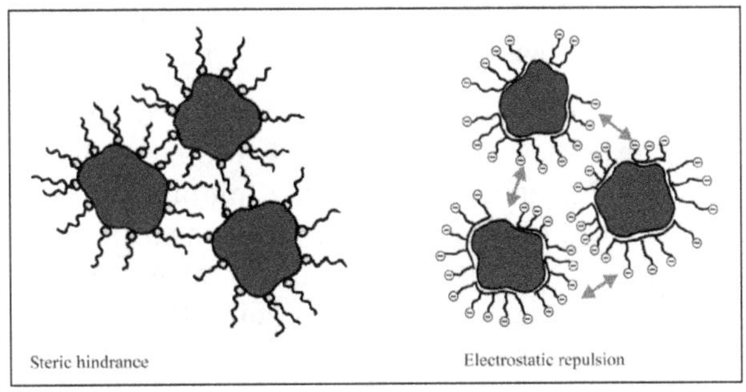

Figure 87: Basic principles of pigment stabilization, source: EVONIK TEGO

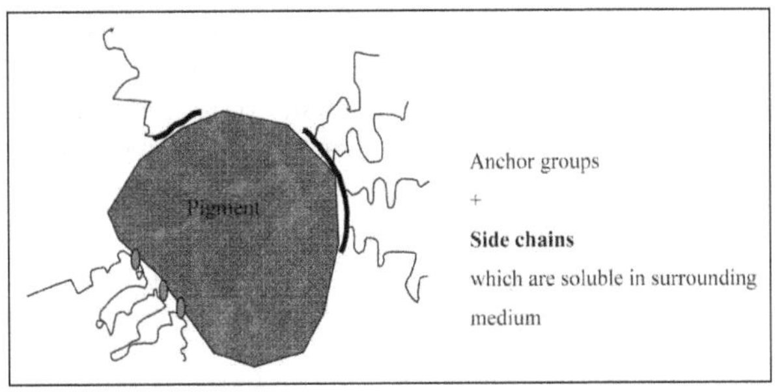

Figure 88: Examples for pigment wetting and dispersing additive structures providing steric hindrance, source: EVONIK TEGO

The mechanism of electrostatic repulsion only works if the wetting and dispersing additive carries ionic groups that dissociate in the coating or ink. While in the case of 100 % liquid RC formulations the insufficient polarity of the medium prevents the dissociation the use of such additives is restricted to water-based UV formulations only.

Therefore, steric hindrance is the stabilizing mechanism of choice in case of 100 % liquid systems. Those additives must bear groups of high pigment affinity that adsorb on the pigment surface [95]. High performance wetting and dispersing additives often contain different groups for pigment interaction in order to achieve good effectiveness on a variety of pigments. Besides aromatic ring structures, carboxylic, amino groups and many others may be implemented. Other parts of the additive molecule are polymer chains that are very compatible with the monomer/oligomer blend and in the final formulation unfold and stretch into the surrounding medium. Interactions of the pigment particles are prevented. Ideally high pigment concentrations can be obtained whilst maintaining low viscosity and newtonian flow, i.e. preventing thixotropy. To achieve good pigment stabilization also upon storage highly polymeric structures are preferably used [96].

Commonly used are:
- special aminofunctional polyesters
- acrylic copolymers
- polyurethanes

Additives solved in organic solvents are undesired. Therefore, tailor-made pigment wetting and dispersing additives with 100 % active matter content, acceptable solubility in monomers, and good compatibilities have been developed and are widely used – especially on organic pigments and carbon blacks. Others preferentially absorb on the surfaces of inorganic pigments. Certainly there are still several traditional solvent-based additives in radiation curing formulations, the trend however is to move away from them towards solvent-free additives.

Note

- Pigment wetting and dispersing additives perform differently depending on the type of pigment used due to the nature their pigment affine groups and that of the pigment surface. For optimal performance different products have to be used for
  - organic and carbon black pigments,
  - inorganic pigments and fillers,
  - matting agents.

  There are single polymeric high performance pigment wetting and dispersing additives available that are suitable for all above mentioned pigment and filler classes. However, some compromises on cost and/or performance must be accepted.
- Pigment wetting and dispersing additives are also used to prevent pigment settling. However they cannot completely overcome the laws of physics. Especially with inorganic pigments and fillers settling may not completely be avoided. This is mainly possible at high viscosities or with shear thinning flow behavior. But even if there is some pigment settling the additive can ensure that the pigments and fillers form only a soft settlement that can be readily stirred up.

3.5.2 Rheological additives

In contrast to solvent- and water-borne formulations there are hardly any specific rheology control additives available for radiation curing formulations. The most common rheological problems here are too high viscosity or undesired shear thinning (pseudo plastic) flow. Suitable pigment wetting and dispersing additives can be used to avoid viscosity build in the presence of pigments, fillers or matting agents (Chapter 3.5.1). But there are no non-VOC additives available that could simply dilute the formulation without affecting film properties. Thus, the desired final viscosity level has to be taken into consideration already when selecting the oligomers and monomers for the formulation.

In several applications a certain rheology is required (e.g., shear thinning to achieve high film thicknesses or to avoid pigment settling). Besides the choice of oligomers and monomers certain additives can be used to control the rheology of RC coatings and inks:

- finely dispersed amorphous silica
- special oligomers respectively additives that convey a certain shear thinning flow behavior to the formulation [97].

Certainly the most common approach to achieve shear thinning flow behavior in RC coatings and inks is the use of amorphous silica obtained through pyrogenic manufacture (fumed silica). The surface of the fumed silica is covered by siloxane groups and hydroxyl groups. The siloxane groups are non-polar and account for the hydrophobic characteristics of the silica surface. The hydrophilic hydroxyl groups are the key to the mechanism through which fumed silica functions [98]. These silicas can be untreated with maximum hydrophilic properties or treated to impart more hydrophobic properties.

When fumed silica is dispersed in a liquid, the silanol groups on adjacent silica aggregates via hydrogen bonds with one another to form a three dimensional network as indicated in Figure 89. The silica network increases the zero shear viscosity. When the system is sheared, the hydrogen bonds between the aggregates are broken and the viscosity decreases. After shearing has ceased, the hydrogen bonds are reformed on standing. The silica network rebuilds and the viscosity increases again. This is called shear thinning or pseudoplastic flow behavior [99]. This property makes the coating thin under the shear forces of application (good flow and

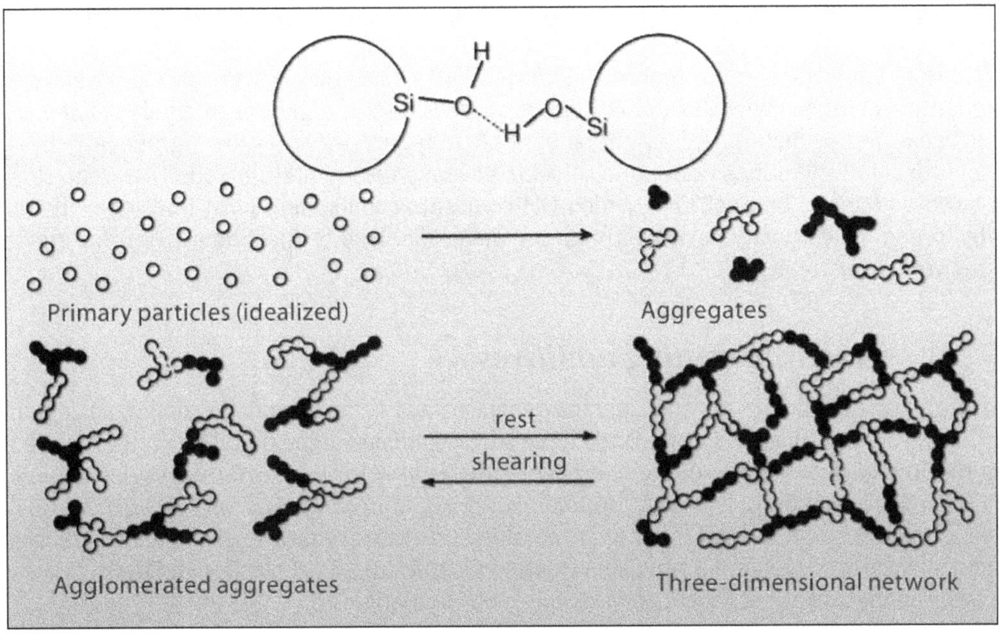

Figure 89: Interaction of silica particles in liquids, source: EVONIK DEGUSSA

leveling), and then thicken on the substrate to resist gravitational sag. When there is a time lag between the end of shear and the rebuild of viscosity the effect is called thixotropy (see Chapter 6.1.2).

By the use of silica it is also possible to adjust a yield point, the amount of shear stress required to initiate flow. Unless the gravitational shear stress on an applied coating exceeds this yield value, no flow will occur and sagging resistance is obtained. The presence of a yield value retards or eliminates settling of pigment particles.

Starting with relatively large particle sizes these silicas are ground in the coating and ink medium to just a few micron average particle size or even to submicron level usually in the presence of pigment wetting and dispersing additives (see Chapter 3.5.1). It is not possible to grind fumed silica down to its primary particle size, which is often just a few nanometres.

The finer the achieved dispersion, the larger the free surface area, and the number of reachable hydroxyl groups for interaction and the stronger the thickening effect. Fine dispersion is also important to avoid possible influence on the optical appearance of the coating, especially gloss.

Note

Silicas are used in the coatings and ink industry for several purposes

- gloss reduction (matting) (see Chapter 3.6.6)
- rheology control
- improvement of scratch resistance

In principle fumed silicas can be used for all three effects. The specific properties however, like surface area, surface treatment, particle size, aggregation behavior and incorporation determine the final effect. Thus, one single product cannot achieve all three effects equally well.

Note

There is often the need to reduce the viscosity of a formulation without adding large quantities of monomers. If the high viscosity is the result of hydrogen bonds between the coating raw materials, it might be possible to reduce the viscosity significantly by adding small quantities of organic solvents or monomers containing one OH-group (if acceptable use 2-butanol). The added OH-containing monomers form intermolecular hydrogen bonds to the polar coatings raw materials. Due to the single OH-group the network is destroyed.

3.5.3 Substrate wetting additives

Wetting can be influenced by the substrate surface and via the formulation of the coating. In principle, a solid surface is well wetted if the surface tension of the liquid is small compared to the free surface energy of the substrate. Substrates with low surface energies include polyethylene, polypropylene and "Teflon", examples of high energy surfaces are metals, metallic oxides and glass. Cleaning of metallic substrates removes grease and increases their surface energy. Corona or plasma treatment of the surface of plastics and films creates a layer on the surface of higher surface energy by oxidation.

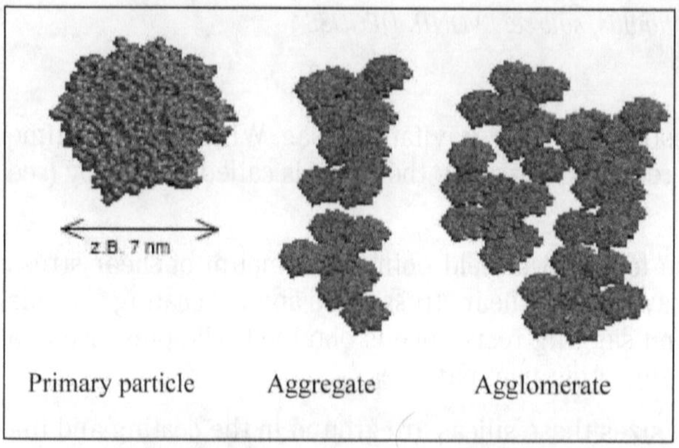

Figure 90: Schematic picture of fumed silica particles, source: EVONIK DEGUSSA

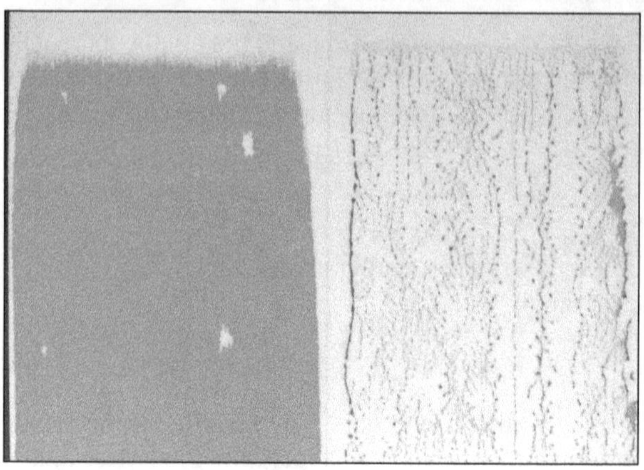

Figure 91: Draw-down of UV ink wetting a low energy surface (silicone release paper), source: EVONIK TEGO left: with substrate wetting additive, right: without additive

The most widely used method for improving wetting is the addition of substrate wetting additives to the liquid phase. These surface active compounds orient themselves preferentially to the liquid-air interphase where they reach a higher concentration than in the bulk phase (Gibbs) (Figure 92) [100]. By reducing the surface tension of the coating or ink they enable the wetting of lower energy surfaces.

In general, substrate-wetting additives act as surfactants in radiation curing formulations. As amphiphilic substances, surfactant molecules combine at least a very polar and a non-polar part within the molecule.

This structure is also responsible for their preferential orientation at the interfaces: The non-polar part is forced out of the volume phase, the more polar part points towards the polar liquid phase. This arrangement of the surfactant molecules along the liquid/air interphase creates the desired reduction of surface tension.

Substrate wetting additives have gained importance in the coating and ink industry with the move towards water-based and RC formulations. Their inherently higher surface tension compared to conventional solvent-based formulations increases substrate wetting problems on many substrates. The following Table 20 shows the surface tension of some liquids and Table 21 of different subtrate materials.

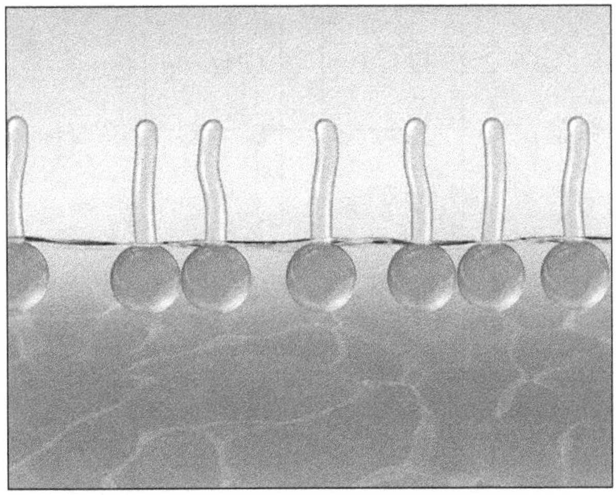

Figure 92: Surfactant orientation at the interface, source: EVONIK TEGO

The most common types of substrate wetting additives used in RC formulations are

- hydrocarbon surfactants,
- silicone-based surfactants,
- fluorinated surfactants.

Hydrocarbon surfactants improve the overall wetting and flow properties of radiation curing formulations without strongly reducing the static surface tension. However, they have an impact on contact angle and dynamic surface tension (measured by maximum bubble pressure technique). Hydrocarbon surfactants can help to improve wetting of substrates like wood, paper, etc. Improved wetting properties can also be obtained by the use of some monomers with long alkyl chains (see Chapter 3.3.3).

When it comes to the wetting of lower surface energy substrates, such as olefinic plastics, poorly degreased metal or fresh litho ink, higher surface activity is required. In RC formu-

Table 20: Surface tension of various liquids [101]

Liquid	Surface tension (mN/m)
water	73
epoxy acrylate in TPGDA	36
butyl glycol	30
toluene	29
isopropanol	22

Table 21: Surface energy of substrate surfaces

Substrate	Surface energy (mN/m)
steel	approx. 50
aluminium	approx. 40
polyester	43 to 45
polyethylene/LD	36
polyethylene/HD	32
polypropylene	30 to 34
paraffin wax	26
PTFE ("Teflon")	20

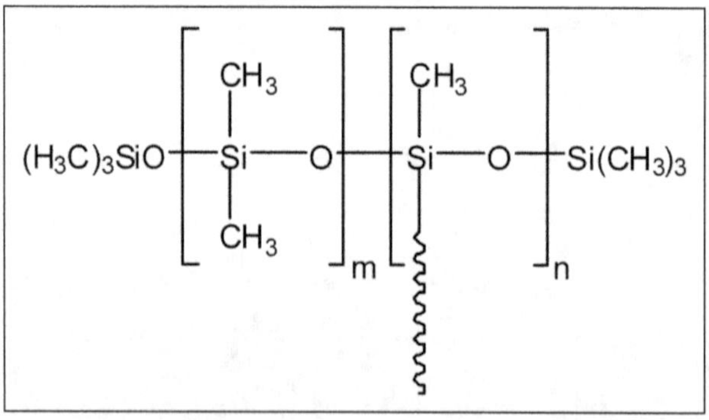

Figure 93: Typical structure of a polyether modified polysiloxane (chain = polyether)

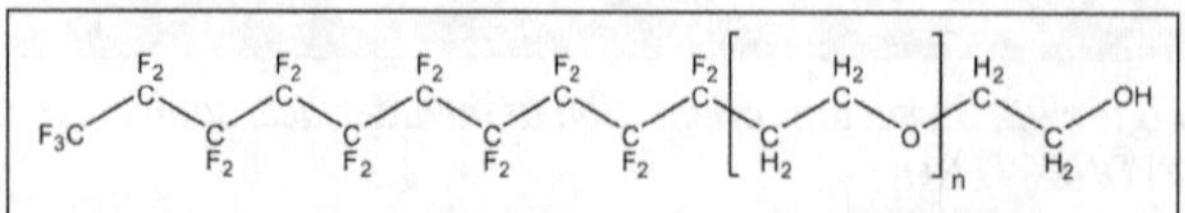

Figure 94: Typical structure of a fluorosurfactant [102]

lations silicone-based surfactants provide a significant reduction of static surface tension. Most compatible and most widely used for this application are polyether modified polysiloxanes with rather hydrophilic polyether chains (Figure 93).

Such kinds of polyether siloxanes need a certain length of siloxane backbone in order to function as a powerful surface tension reducer in radiation curing formulations. In general, the length of the siloxane backbone (i.e. m+n+2) varies between 5 and 50. The polyether modification can be strictly ethylene oxide-based, but may also consist of a blend of ethylene oxide and propylene oxide. In the latter case the ethylene oxide is most often dominating the chain. The siloxane backbone forms the very non-polar chain of the surfactant, and the polyether chains represent the polar sections. When choosing a polyether siloxane as substrate wetting additive it should be considered that shorter chain products show milder surface tension reduction, but stay recoatable, reprintable and glueable. Polyether siloxanes with a siloxane backbone higher than 10 siloxane (Si-O) units form an additive layer on top of the cured coating or ink that may interfere with recoatability but improve the slip properties. If a combination of powerful substrate wetting and slip is desired, those polyether siloxanes are the ideal candidates (see Chapter 3.5.6).

Even more surface active than many silicone-based additives are fluorinated surfactants. In these surfactants the non-polar group consists of a perfluorinated alkyl chain, the polar part of the molecule plays the solubilizing role in the formulations and may consist of different structures (Figure 94). It needs to be sufficiently polar in order to achieve the desired

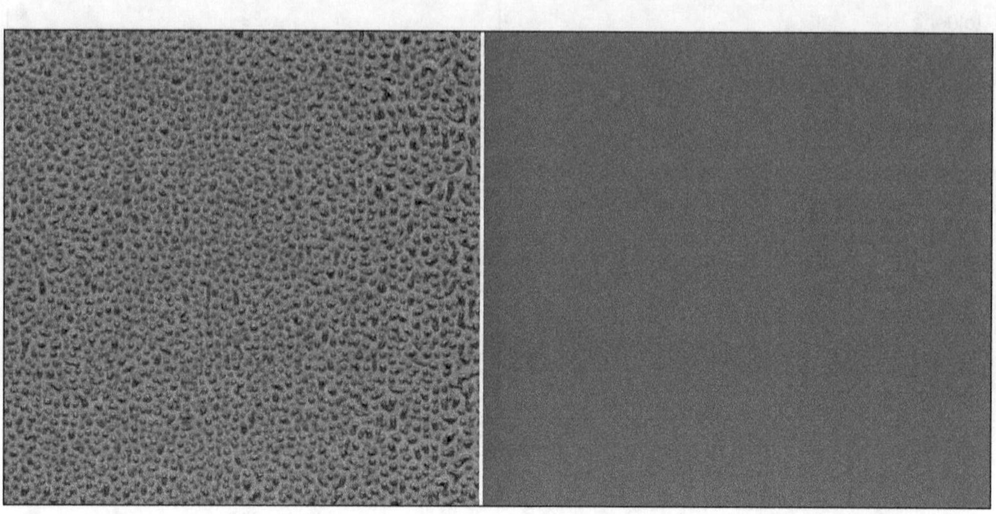

Figure 95: Microfoam in a screen-printed UV varnish left: without deaerator, right: with deaerator

Table 22: Static surface tension of surfactant solutions in TPGDA (0.3%, 25 °C)

Surfactant	Surface tension (mN/m)
hydrocarbon surfactant	32.5
short chain polyether siloxane	27
longer chain polyether siloxane	24
non-ionic fluorosurfactant	22

solubility and can be of ionic or non-ionic nature. In radiation curing formulations non-ionic head groups are preferably used.

The cohesive forces between the perfluoro alkyl groups are particularly low and the surface tension of the formulation is correspondingly low. Perfluoro alkyl chains however show a very limited compatibility with radiation curing formulations which limit their use. Apart from this special attention should be paid to potential foam stabilization. They are highly suitable to prevent craters and fight contaminations. Due to their negative aspects of the environment more and more raw material producer as well as coating/ink manufacturer avoid the use of fluorinated raw materials. Table 22 shows the static surface tension of surfactant solutions in monomer. Table 24 lists main properties of substrate wetting additives.

Note

When selecting a substrate wetting additive one of the crucial factors often is the foam stabilization that may go along with the use of such a product. Because of this it is highly recommendable to start off with the least foamy product for testing. Tendentially the highest risk of foam stabilization is given when using fluoro-surfactants followed by highly compatible, highly active polyether siloxanes.

3.5.4 *Defoamers and deaerators*

Foam is a major problem creator in many coating and ink applications. Not only does it increase the volume of the coating or ink and lead to surface defects in the cured film. Fine foam also increases the viscosity of the formulation and may make the application impossible. It also reduces gloss and transparency and reduces the protective effect of a coating.

Foam is a stable distribution of small gas bubbles (usually air bubbles) in a liquid coating or printing ink. Air incorporation can occur:

- during coating or ink manufacture (grinding or mixing),
- by application processes such as rolling, dipping, spraying, airless-spraying and printing,
- after application when wetting porous substrates such as wood [103].

The presence of surfactants in the formulation may stabilize foam strongly and the inherent viscosity of RC formulations slows down the air release. This in combination with the cure immediately after application of RC formulations makes foaming a very critical issue. In literature a distinction is made between microfoam and macrofoam, between very small air bubbles surrounded by a surfactant layer and high volume foam forming lamellas with surfactant double layers [104]. In the case of radiation curing formulations however, the dominating appearance of foam is microfoam, sometimes combined with some foam bubbles on top of a vessel of paint film (Figure 96).

There are various ways of combating foam in coatings and printing inks. Avoiding foamy surfactants, choice of low-foam raw materials or adjustment of viscosity to an optimum value can markedly reduce micro-foam problems. The manufacturing and application processes should also be organized so that only a minimum of air is incorporated into the coating material. Since items such as raw material and methods of manufacturing and application are usually specified the room for manoeuvre is very limited. Addition of additives such as deaerators and defoamers is the simplest way to inhibit and remove foam, often the only way.

There are many complex theories discussed in literature describing how additives manage to remove microfoam and make macro bubbles burst. Deaerators are often described as spreading around the fine bubbles, forming a thin liquid layer and preventing stabilization by surfactants. Additionally or alternatively they may speed up the dissolution of the air bubble in the coating. Defoamers are often described as insoluble droplets entering the foam lamella and making it burst [105] (Figure 97 and Figure 98).

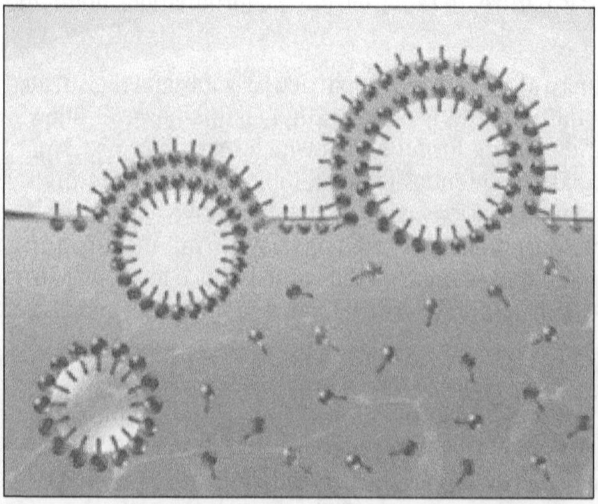

Figure 96: Formation of micro and macro foam in a coating, source EVONIK TEGO

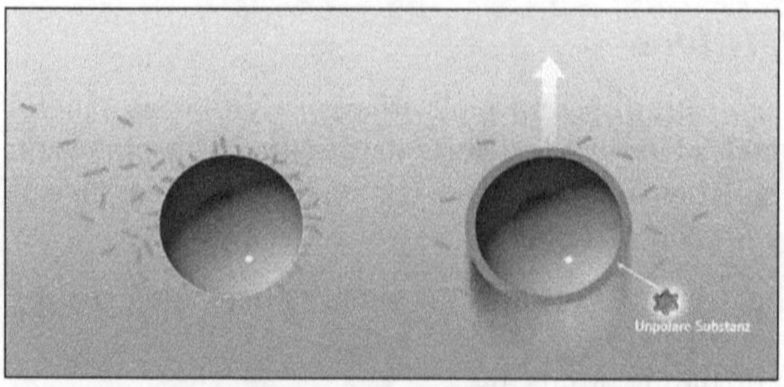

Figure 97: Dearation mechanism, displacement of surfactants by deaerator, source EVONIK TEGO

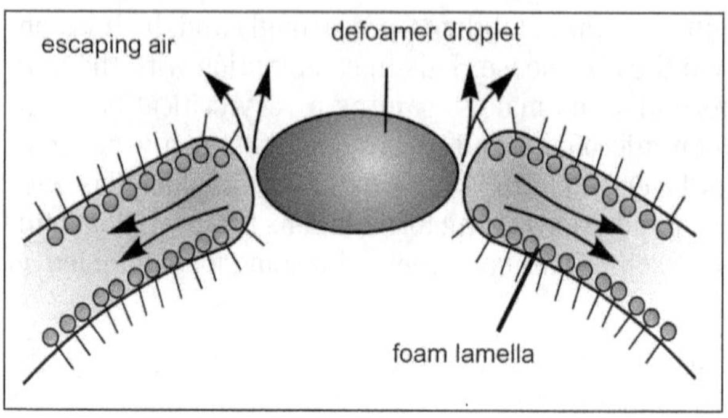

Figure 98: Defoaming mechanism, source EVONIK TEGO

In practice this distinction is not always possible. In any case there are certain properties required to make a substance work as defoamer or deaerator in liquid coatings and inks: it needs to be liquid, display a certain incompatibility or (partial) insolubility, has a rather low surface tension and high spreading activity. Effective deaerators have targeted incompatibility with the coating system. However it is necessary to find a good balance between effectiveness and compatibility: if the compound is too compatible and soluble, it will not exhibit deaerating activity, but if it is too incompatible, there will be defects such as craters, fish-eyes or cloudiness [106].

There is a wide range of chemical compounds which show targeted incompatibility in radiation curing formulations. Each class of substances has distinct advantages and disadvantages in application. The absence of organic solvents and the resulting low solvating power makes the choice of deaerators and defoamers for RC formulations especially critical. Additive suppliers have developed specific additives for this application which display the best possible balance between targeted incompatibility and the avoidance of surface defects [107].

Table 23: Most commonly used deaerators/defoamers

Substance class	Pros	Cons
dimethyl polysiloxanes (silicone oils)	utmost effectiveness	high risk of turbidity and surface defects like craters, poor recoatability, risk of separation
organically modified polysiloxanes	very high effectiveness, very versatile	some risk of incompatibility, may interfere with recoatability
fluorosilicones	high effectiveness, especially against foam stabilized by surfactants	risk of incompatibility, only provided as solutions in organic solvents
special organic polymers	limited effectiveness especially in pigmented formulations and fast process (e.g. rotary screen printing)	excellent recoatability/reprintability, good compatibility in clears

Due to the wide range of raw materials used in RC coatings and inks no single additive is able to prevent foaming problems in all applications and formulations. UV clears require utmost compatibility whereas pigmented formulations can tolerate and actually need a higher level of incompatibility. Table 23 lists the commonly used deaerators and defoamers.

3.5.5 *Antioxidants and formulation stabilizers*

Stability of an acrylate formulation during oligomer manufacture or formulation storage can be affected by a number of parameters such as temperature, oxygen, light, pigments or fillers, impurities such as metal ions, and the presence of amine or thiols. Unwanted radical formation leads to premature polymerization of the oligomer or to formation of hydroperoxide radicals. Stabilizers are substances which are added to the system in small amounts during manufacture and storage to prevent auto-oxidation of the acrylates and thus to

- reduce the risk of early polymerization during processing, and
- increase the shelf life of the formulation.

They perform this function by reacting with oxygen and carbon-centered radicals, preventing them from initiating radical polymerization. For formulations stored in the presence of air, stabilizers are usually chosen from a class of hydroxyaromatic compounds: phenols and quinones [108] are common stabilizers added to acrylate formulations (Figure 99) but their performance deteriorates as soon as the formulation composition becomes more complex (i.e., in the presence of initiators, amines, etc). In addition, some of them show poor toxicological profiles; hydroquinone, for example, is possibly carcinogenic. The stabilization process with BHT (2,6-di-tert-butyl-4-methylphenol) is illustrated in Figure 100.

For storage under inert conditions, phenothiazine [109] or N-oxyls (e.g. TEMPO) are used to prevent premature polymerization: however, they affect cure efficiency and impart a strong discoloration to the film after UV exposure.

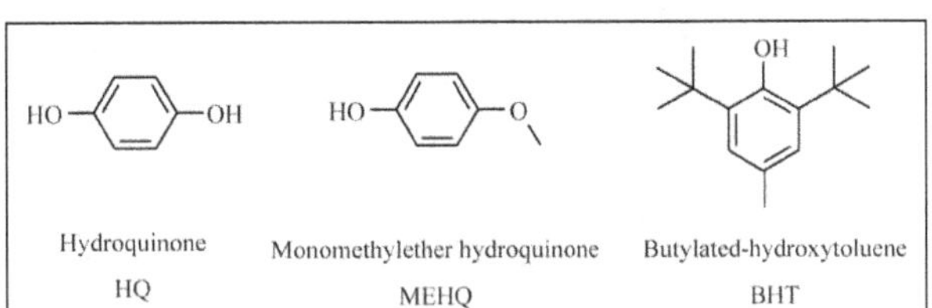

Figure 99: Examples of stabilizers

N-nitrosophenylhydroxylamine salts and dithiocarbamate salts (Figure 101) can also be used as stabilizers but again show poor toxicological profiles.

New in-can stabilizers, one nitroxyl- and the other phenol-based,

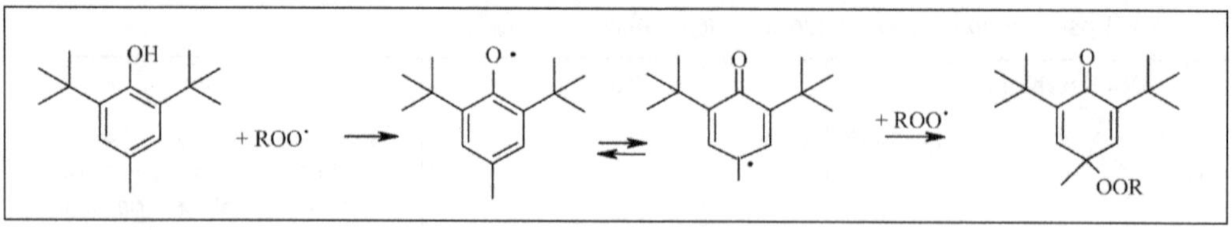

Figure 100: Stabilization mechanism with BHT

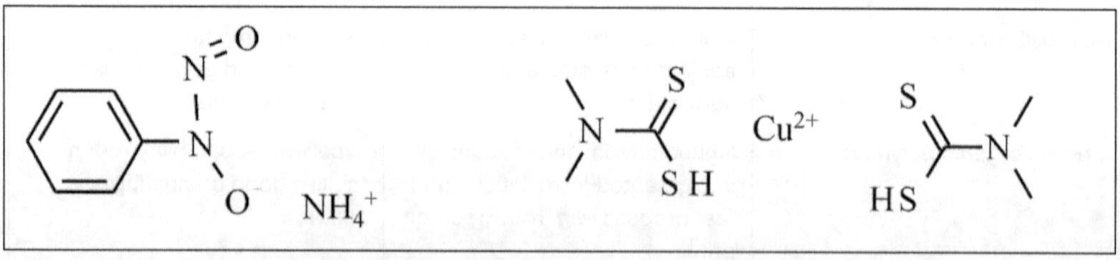

Figure 101: Structures of N-nitrosophenylhydroxylamine salts and dithiocarbamate salts

for UV curable formulations have recently been introduced on the market for storage under air or an inert atmosphere; these provide excellent stability even when incorporated in formulations of complex composition and show good toxicological profiles. Moreover, they do not affect cure performance when used at low levels, and cause hardly any discoloration of the film [110].

3.5.6 Surface control additives: flow, leveling, slip, scratch resistance

The term surface control comprises a variety of effects that finally determine the appearance and surface properties of the cured films (Figure 102).

Flow stands for the formation of a closed film with a rather homogeneous layer thickness directly after application. The better the flow the better are irregularities of the substrate covered from the film.

Leveling stands for the smoothing of fine irregularities in the upper area of the coatings layer. A poor leveling results for example is an orange peel effect.

Slip is the common term for creating a slippery surface or in other words reducing the coefficient of friction (COF). A good slip improves properties like scratch resistance.

3.5.6.1 Flow and leveling additives

The terms flow and leveling are often used together or even interchangeably. But as can be seen in Figure 102 and from the definitions their mode of action is different. Flow additives in RC formulations often work via their surface activity. Many products can actually be considered substrate wetting additives and flow additives as well. However, in the case of flow additives the strong reduction of surface tension (like in the case of wetting additives) is of less importance, the uniformity of surface tension of greater importance.

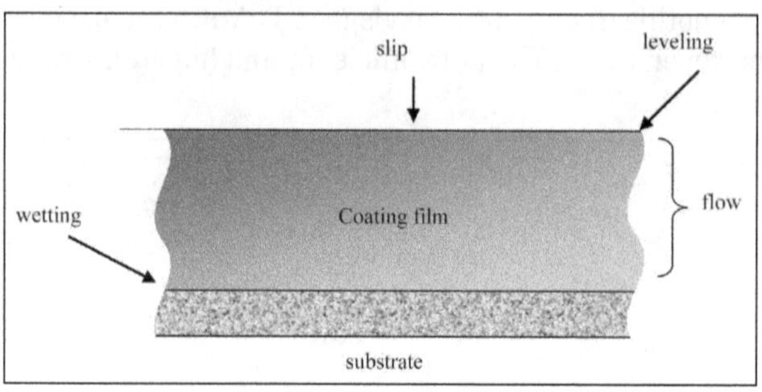

Figure 102: Distribution of effects throughout a coating film

Table 24: Main properties of substrate wetting, flow and leveling additives in RC formulations

	Substrate wetting additive	Flow additives	Leveling additives
Surface tension reduction	strong	medium	mild to none
Solubility in formulation	good	good	good to acceptable
Potential defoaming effect	hardly ever	seldom	possible
Potential foam stabilization	often	possible	seldom
Use of siloxane-based structures	often	relatively often	seldom
Use of organic polymers	hardly ever	sometimes	often

Consequently leveling additives are often silicone-free surfactants or polymers (particularly polyacrylates). The following Table 24 lists the main properties of substrate wetting, flow and leveling additives to make the distinction easier.

3.5.6.2 Waxes – slip additives

In radiation curing coatings and inks the slip properties and scratch resistance are often improved by the use of waxes [111, 112]. Now the word wax is a very general term and comprises a variety of different substances, all of which have a few properties in common [113]:

- solid at 20 °C, varying from soft/plastic to brittle/hard
- melting point > 40 °C without decomposing
- relatively low viscosity slightly above the melting point
- transparency to opaque, but not glass-like

Such materials are available from natural sources, such as Carnauba, Lanolin, beeswax, etc. More common for the use in coatings and inks however are synthetic waxes based on

- polyolefines (PE and PP)
- polyamide
- PTFE
- paraffin (Fischer Tropsch)

The different products are shown in Figure 104.

Most commonly used in radiation curing coatings and inks are micronized waxes with an average particle size between 3 and 10 μm. In principle they can be incorporated into the formulations without pre-dispersion by stirring if sufficient shear forces can be applied. In the case of liquid RC formulations often a pre-dispersion in monomer with the help of a wetting and dispersing additive is prepared. Common addition levels for micronized waxes are 0.5 to 5 % solid wax on total formulation.

The mode of action is based on two principal mechanisms: By stretching out of the surface the wax particles provide a mechanical protection. In the case of an object scratching along the surface or another coating/ink layer gliding along, the wax particles keep the films apart from each other so that they cannot be damaged.

The second effect of waxes is closely related to their melting point. The tips of the wax particles rub against each other or against the other surface in case of a mechanical attack. The result is initially a very high frictional force that is set free at the tip of the wax particle as heat. Depending on the friction and the melting point of the wax the wax particle melts partially and the liquid acts as a kind of lubricant by hydrodynamic lubrication. The surfaces glide along each other readily.

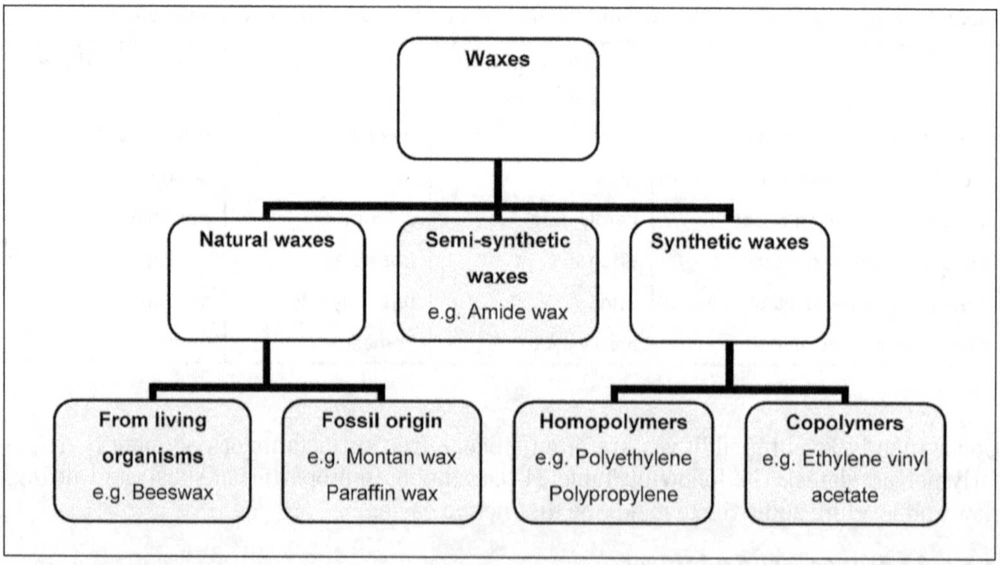

Figure 104: Different kinds of waxes, source EVONIK DEGUSSA

Rather soft waxes (i.e. with a rather low melting point) contribute strongly to the slipperiness of a coating surface. Very soft waxes may convey a nearly smeary feel and cannot provide a high degree of mechanical protection. Harder waxes create a dryer feel, but convey excellent scratch and rub resistance. Only PTFE waxes are rather hard, but due to their low intermolecular forces create a very low coefficient of friction (COF).

It is also possible to create a certain anti-slip effect by the use of waxes. Highly suitable for this application are very hard polypropylene waxes that create a very high COF, but still give mechanical protection to the cured film. To improve properties like incorporation many

Table 25: Classification of waxes

Wax type	Description	Effect
PE	obtained by polymerization of ethylene monomers low density polyethylene waxes display more toughness, better scuff/mar resistance high density polyethylene waxes contribute to better slip properties and abrasion resistance	slip abrasion resistance anti-blocking
PP	obtained by polymerization of propylene monomers good film toughness but poor slip and lubricity properties may be used to convey anti-slip properties	anti-slip anti-blocking abrasion resistance matting
PTFE	obtained by polymerization of C_2F_4, may also be used to prepare PTFE-modified PE-waxes monomers very hard, high melting point give the lowest known coefficient of friction excellent slip & lubricity, anti-blocking and abrasion resistance	slip anti-blocking abrasion resistance
Paraffin	generally obtained by the Fischer-Tropsch process, mainly composed of straight alkane chains very low polarity, little use in RC formulations	anti-blocking water repellence slip

Table 26: Comparison of different slip additives

	Waxes	Siloxane-based additive	Acrylated siloxane-based additive
COF reduction	moderate to strong	strong to very strong	strong to very strong
Improved abrasion resistance	very good	moderate	moderate
Anti-slip possible	yes	no	no
Recoatability	moderate to good	critical	very critical
Gloss and optical clarity	reduced	unchanged, depending on selection	depending on selection
Release-properties	moderate	strong	very strong
Additional wetting and flow promotion	no	yes/possible	yes/possible

types of waxes are modified further, e.g., by oxidation reactions. Table 25 classifies some wax types and Table 26 compares different types of slip additives.

In addition to the above described improvement of slip, scratch resistance and anti-blocking waxes are also used to

- improve water repellence (because of their hydrophobic nature),
- matt radiation cured coatings (due to their solid nature and particle size). Addition levels may then reach up to 10 % (> matting).
- improve sandability (in case of amide waxes)

3.5.6.3 Siloxane-based slip additives

A strong slipperiness (COF-reduction) can be achieved by the use of siloxane-based slip additives – also called glide additives. Like substrate wetting additives they are surface-active but have a higher molecular weight and a strong tendency to orient at the film/air interface. During curing, siloxane-based additives form a thin lubricating film on top of the surface and thus improve slip properties and scratch resistance [114].

For compatibility reasons most formulators refrain from using polydimethyl siloxanes (silicone oils) as glide agents in RC formulations. Preferably organo modified polysiloxanes are used, that can be adapted to the required property profiles by varying their siloxane chain length, the degree of modification and the nature of the modifying organic chain. It is very important to mention that these products do not cause defects like craters as known from silicone oils. Longer siloxane backbones ensure stronger slip and release properties, while their specific incompatibility increases. A higher degree of modification improves solubility, compatibility and recoatability – at the same time specific surface related effects such as release and anti blocking properties are reduced. Linking polyethers to the siloxane molecules dramatically improves the polarity of the additives, so that compatibility with the rather polar formulations is optimized and application becomes even possible in aqueous systems [115]. Figure 105 shows the effect of improved solubility by polyether modification.

Hydrophilic polyether siloxanes with a medium siloxane chain length are often used, which can really be considered multifunctional. They combine effective substrate wetting, flow promotion and slip in one additive. They are widely used in overprint varnishes or wood top coatings. Limited is their use in applications where recoating without sanding or reprinting/gluing in an offline process is required.

A general rule says: the shorter the siloxane chain, the weaker the slip and the better the recoatability. Therefore, if slip and recoatability are required compromises must be made.

The slip effect is quantified by measuring the force required when a standardized object of defined weight is pulled across the cured formulation at a specified speed. The lower the slip value the more effective the additive. It can be seen from Figure 106 that different types of slip additives perform diffently.

The concentration of long chain polysiloxane additives at the air-interface can be used to obtain even more extreme effects, such as release. Siloxane-based additives with long siloxane backbones and relatively high incompatibility are used to maximize the orientation of the siloxane and the resulting release effect. These effects can be especially durable in the case of a crosslinkable nature of the additive.

3.5.6.4 Crosslinkable siloxane-based additives (silicone acrylates)

The ability to crosslink is especially interesting for additives that act at the formulation to air interface i.e. for surface control and release additives. The incorporation into the polymeric network of the UV formulation effectively prevents the transfer of these additives to other surfaces, and prevents de-migration when in contact with liquids, enabling a longer lasting effect to be obtained. Therefore, acrylate functional additives appear to be the appropriate approach (Figure 107). Products of similar chemical structure are also often used as oligomers in release coatings (see Chapter 3.3.2.5).

Figure 105: 0.3 % silicone oil (left) and polyether siloxane (right) in UV clear, source: EVONIK TEGO

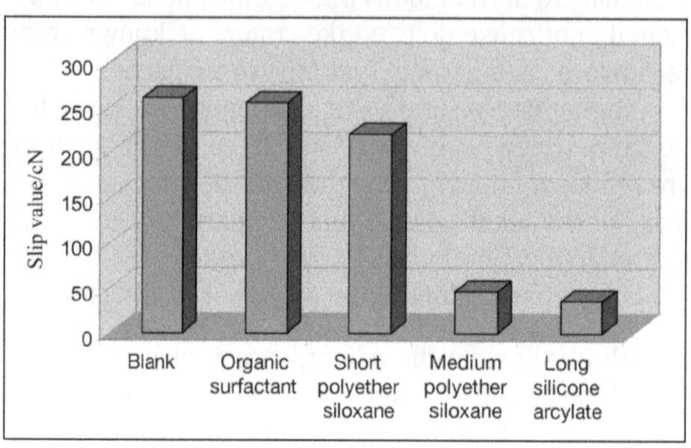

Figure 106: Slip of different additives in epoxy acrylate UV varnish

By varying the structural parameters of silicone acrylates a variety of different property profiles can be obtained. In addition to their slip properties and release effect significant differences can be obtained with regards to deaeration properties, flow and leveling, foam behavior. To ensure good compatibility in RC varnishes and inks organomodified silicone acrylates are preferably used.

The greatest differences between the different goups of silicone-based additives can be observed when measuring the release properties. An adhesive tape is applied to the cured coating under defined conditions and then peeled off at a constant speed and the required force measured as release value. The more effective the release addi-

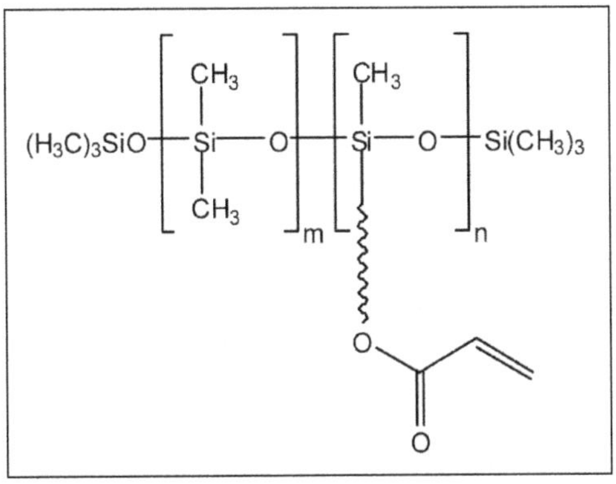

Figure 107: Typical structure of an organo modified silicone acrylate, chain = alkyl

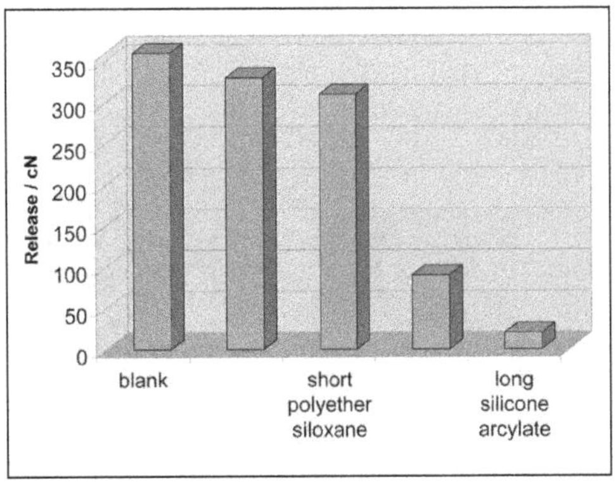

Figure 108: Release effect of different additives in epoxy acrylate UV varnish

tive, the lower is the required force. The efficiency of different types of slip additives is shown in Figure 108.

Without any doubt it is possible and often advantageous to combine waxes and siloxane-based additives to control the slip properties and scratch resistance of a RC coating or ink. In Table 26 a comparison of different slip additives is shown.

Note

It is not easy to adjust a defined slip level that stays stable over time. For stable slip properties it is important that the surface of the cured film is fully covered by the slip additive. Therefore, it is recommendable to use more than the following Figure 109 indicates. The most difficult task is to create a high or medium coefficient of friction (COF) on a stable basis. Just to reduce the amount of highly active slip additive to a very low level may provide inconsistent COF readings. Minor differences in application and cure conditions would change the slip properties too strongly. Here it is recommended to use a milder (weaker) slip additive at higher dosage. It may be beneficial to make combinations with waxes (see Chapter 2.5.6.1.) or - in case of high COF coatings or inks - to use waxes alone.

3.5.7 Matting agents

Radiation cured varnishes and inks are known to display excellent gloss - and efficient matting can be a real challenge for the formulator. Extremely even and smooth surfaces display very high gloss because they provide a very exact reflection of light. Low gloss goes along with a more diffuse light scattering.

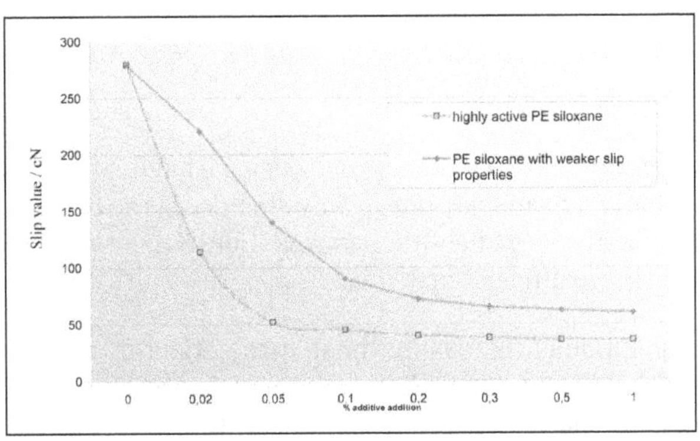

Figure 109: Slip properties of UV clear with two different PE siloxanes (Slip value in cN)

To make a coating or ink surface appear matt it is necessary to create a certain micro roughness

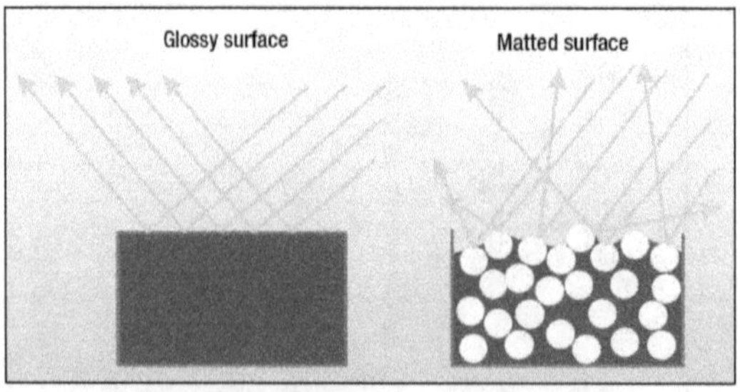

Figure 110: Surface matting by micro-roughness [116], source: EVONIK DEGUSSA

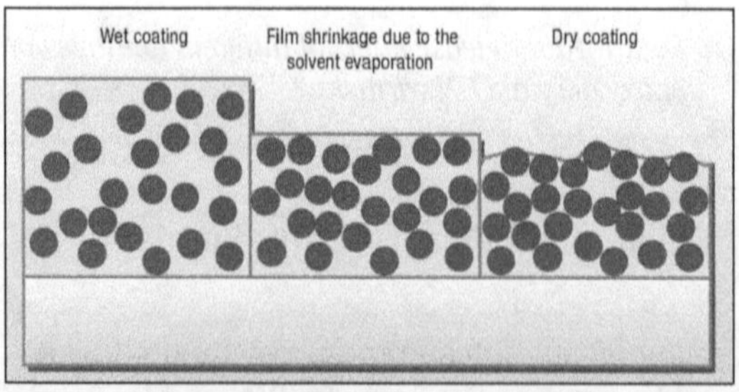

Figure 111: Matting effect due to volume shrinkage [117], source: EVONIK DEGUSSA

Figure 112: REM photograph of UV cured clear coating film (100 % solids, matted with 8 weight% silica), source: EVONIK DEGUSSA

at the interface coating/air. This can be achieved by coarse solid particles that create the desired structure (Figure 110).

In solvent- and water-based coating systems this effect is achieved by rather large particles with an average particle size of 3 to 14 μm. During drying the solvent evaporates and the applied film shrinks. As a result the film layer thickness is reduced and the coarse particle determine the surface structure creating the desired micro-roughness (Figure 111).

In the case of clear coatings this matting effect is desired without sacrificing any optical properties, especially the transparency. Consequently the matting particles used need to have a refractive index close to the one of the cured coating film. The closer the refractive indexes are the more transparent appears the final film. Various kinds of solids can be used to convey a matt effect to coatings and printing inks. The most commonly used kinds of matting agent are:

- synthetic silicas
- waxes
- coarse fillers and pigments (in pigmented formulations)

The most efficient and most widely used matting agents are synthetic silicas manufactured by the process of

- silica gels
- thermal silica generation
- precipitation

For radiation curing formulations the available classes of matting agents basically are the same as in conventional coatings and inks. But there is one decisive factor which makes matting much more difficult: RC formulations display hardly any shrink. So the basic mechanism that makes matting agents so effective does not function well. Consequently higher levels of matting agents are needed to achieve efficient matting.

The option of increasing the matting effect by making the silica float to the surface of the coating before curing has not proven successful. In the following REM photograph of a UV coating matted by synthetic silica (Figure 112) it is clearly visible that the silica particles are distributed evenly throughout the film. The above mentioned micro-roughness is clearly visible as well and the film appears matt.

Today it is state-of-the art to use silica types with a wax treatment. The surface modification reduces the settling tendency of the silica particles in the formulation upon storage. With regards to the ideal particle size of matting agents latest investigations have shown that different rules apply in radiation curing compared to conventional formulations.

UV coatings with rather high film thickness (60 to 100 µm) can be matted most effectively by silicas with rather small average particle size of 4 to 6 µm. Rather thin films (below 25 µm) are matted best by coarse silica particles (8 to 14 µm) [118]. Depending on the formulation and application heterogeneous particle size distributions may offer distinct advantages over a homogeneous size distribution. The relatively high dosages (commonly between 5 and 15 weight%) in RC formulations often lead to rheology problems (pseudoplasticity or thixotropy) [119].

Today special grades of silica-based matting agents are available, in which the silica particle carries a special surface treatment that offers benefits with regards to rheological performance. New generations of matting agents based on synthetic silicas treated with polysiloxane derivatives have recently been launched. They could provide higher matting efficiency combined with superior rheological performance.

Waxes provide a weaker matting effect because they have a different particle morphology compared to silicas. They are less aggregated and much more compact. At the same time waxes provide many other beneficial effects like improved slip, scratch resistance, etc. (see Chapter 3.5.6.1). For matting purposes only micronized polyolefin waxes are commonly used. Combinations of waxes and silicas are used in order to achieve specific haptical properties (touch and handle).

Note

Precipitated and thermal silicas can be easily incorporated into radiation curing formulations by high speed mixing without pre-dispersion. Grinding and excessive shear forces may not be applied since they would destroy the particle aggregates, reduce matting efficiency and increase viscosity.

When preparing matting agent pre-dispersions (e.g., in monomer) special attention has to be paid to the choice of wetting and dispersing additive because it may interfere with matting efficiency (see Chapter 3.5.1).

3.5.8 Additives to improve adhesion

As already described in Chapter 3.3.1 the adhesion of a radiation curing formulation depends on several different parameters. Assuming good wetting is already ensured (substrate wetting) the adhesion can be improved by 3 principally different approaches:

- by reducing film tension and volume shrink,
- by increasing the interactions of the cured polymer with the substrate,
- by chemically bonding the cured coating with the substrate.

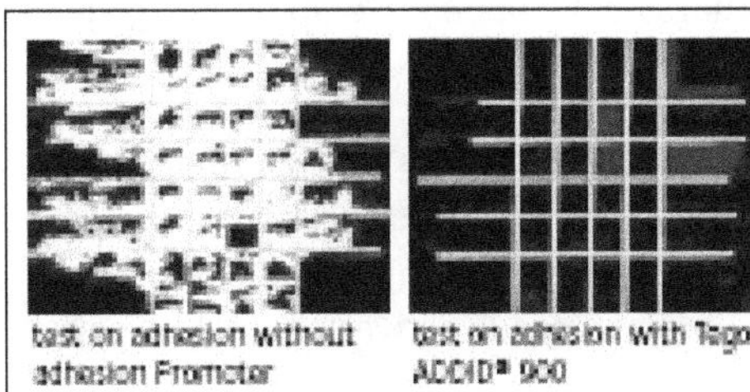

Figure 113: Effect of an adhesion promoter

The different kinds of additives and co-binders used make use of one or two of the above mentioned effects (Figure 113).

Thermoplastic co-binders

Thermoplastic co-binders reduce the film tension and volume shrinkage in RC coa-

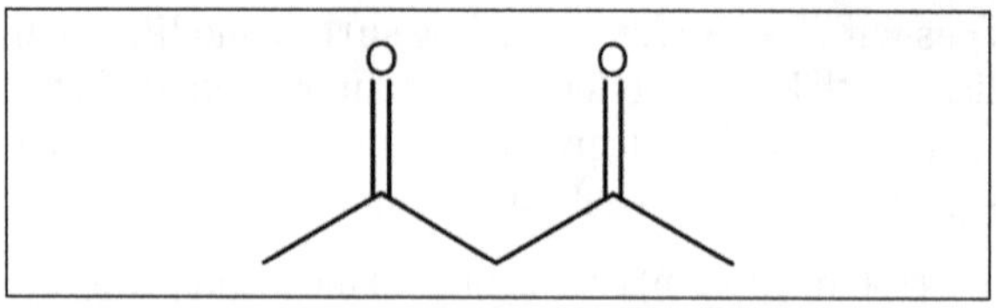

Figure 114: Acetyl acetone (IUPAC-Name: Pentan-2,4-dion)

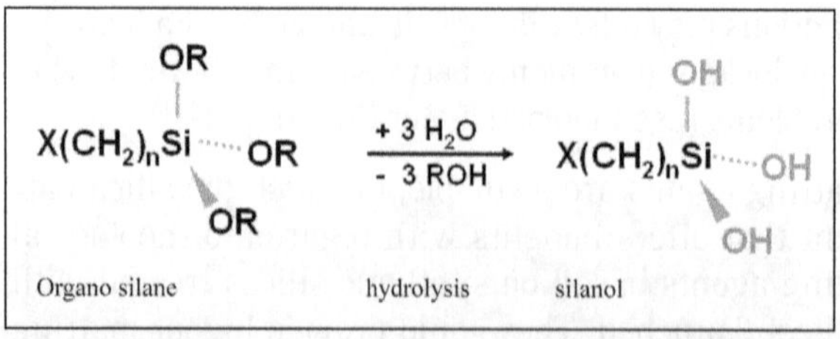

Figure 115: Activation of organo siloxane by water

tings and inks. Additionally they may carry functional groups that increase the interaction with the substrate. Details about the mechanism of volume shrinkage and how to reduce it are discussed in Chapter 3.3.1 and 3.3.2.9.

Some of the commonly used products in the industry are polymeric resins with molecular weights between 1,000 and 10,000 g/mol. These may be based on e.g.,

- ketone/aldehyde resins
- special modified polyesters [120]
- polyurethanes

Besides adhesion promotion to several substrates like metal or plastics these co-binders may have a positive effect on flow, gloss and pigment wetting. It has been reported that hydrogenated ketone resins are suitable as adhesion promoter to substitute genotoxic titanates (see below).

Special dual cure oligomers

Those oligomers carry acrylic and isocyanate functionality at the same time. During UV cure the acrylate group crosslinks within the polymer matrix. Given the fact that there are no or very few hydroxyl groups on the basic raw materials their isocyanate groups can then react with OH-groups from the substrate. So the adhesion to wood and other OH-bearing substrates is improved (see Chapter 3.4.3).

Organic titanates and titanium chelate complexes

Organic titanates and titanium chelate complexes have been very popular adhesion promoters mainly in solvent-based printing inks for many years. The most common structure used to improve adhesion of inks to pre-treated film like PP and PE is titanium acetyl acetonate (Figure 114).

These chelate complexes react with the binder and potentially also with functional groups on the substrate. Besides adhesion promotion their internal "resin crosslinkage" leads to improved heat stability and solvent-resistance, etc. [121]. It has been reported in patents that such titanium chelate complexes are also functional in radiation curing formulations [122]. Their use requires the presence of reactive groups (ideally hydroxyl groups) on the oligomers or monomers. Latest news is that pentane-2,4-dione = acetyl acetone is genotoxic. Even though it is not used on its own in the formulation these news let us expect that the use of acetyl acetonates in the industry will go down dramatically. Therefore, some new titanates have been developed using special resins as ligand instead of acetylacetone.

Organo silanes

Organo silanes are well known adhesion promoters in the coating and ink industry. They display excellent adhesion properties especially on glass and metal by chemically bonding the cured coating to the substrate. The truly active compound for adhesion promotion is the silanol. A silane needs activation (hydrolysation) by water in order to form a silanol (Figure 115).

The silanol groups can react with the substrate's and, possibly, the binder's free OH-groups. In addition to the active silanol groups, the adhesion promoters also carry other functional groups, such as amino or epoxy functionality. These, in turn, can react with the binder present in the formulation.

Note

For the use in RC formulations the need for water as activator appears a challenge. Certainly there is some humidity in the air or at the substrate. But the addition of minor amounts (e.g., 1 %) of water to the silane-containing UV formulation can improve the resulting adhesion on glass or metal significantly.

3.6 Colorants

3.6.1 General introduction

The technology of radiation curing is well established for clear coat applications. This technology is unique due to the combination of features like curing within a fraction of a second, the possibility to formulate coatings without organic solvent and the outstanding quality of coatings thus obtained. Nonetheless, in order to enter into further market segments, some restrictions have to be overcome. One of the most demanding challenges is the UV curing of pigmented coatings. An intense collaboration of raw material suppliers, coating producers and end users over the last years has allowed extending the UV technology also to include the field of colored UV coatings, albeit some issues still need to be resolved [123].

For coloring, two types of colorants can be used: pigments and dyes. By definition, pigments are not soluble in the application medium, whereas dyes are. As a consequence for pigments the particle size and crystal modification are important whereas, for dyes the molecular structure and its chromophore are key to the color produced.

For curing with UV radiation, light is used to trigger the polymerization process by the excitation of a photoinitiator. As a consequence, it is necessary to account for the optical properties of all components of the UV curable formulation, since they may compete with the photoinitiator for the incident light. Dyestuffs also have a major influence on the curing process in colored formulations. The optical properties of dyes are determined by the absorption characteristics, as absorption is the only interaction of the dye with light. In case of pigments, in addition to the absorption process, other optical properties such as scattering and reflection occur and have a significant influence on the curing process.

Thus, in a UV curable pigmented coating the following effects have to be considered:

- **Reflection**: When the incident light hits an interface between two materials of different refraction indexes, its direction is changed and it leaves the interface under the same angle as it had hit the interface. Since the light is usually not completely reflected, this results both in a change of direction and intensity of the reflected light. This effect occurs either at the surface coating/air interface without

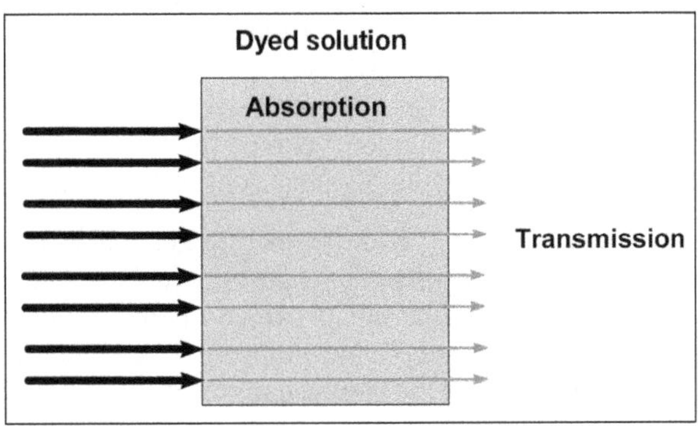

Figure 116: Schematic description of optical effects: absorption and transmission, occurring in a dyed solution

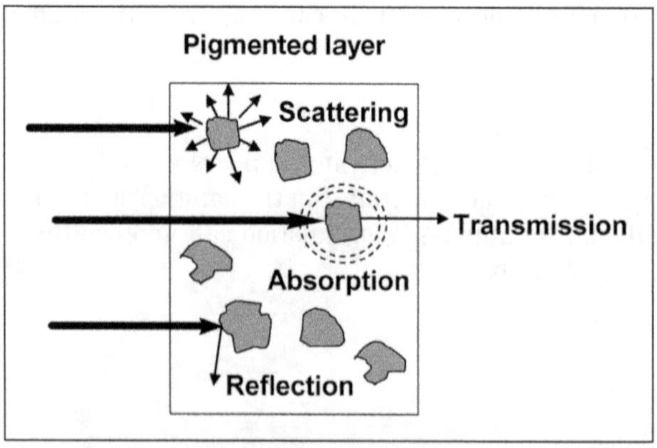

Figure 117: Schematic description of optical effects: scattering, absorption, reflection and transmission, occurring in a pigmented layer

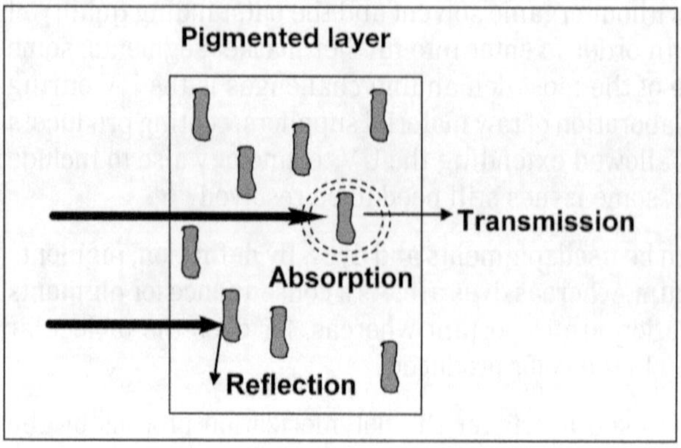

Figure 118: Optical effects of opaque platelets shape pigments

penetration of light into the medium, or after penetration into the lacquer at the surface of coating components such as pigment particles.

- **Scattering**: The nonspecific direction of radiation is deviated into different directions while hitting a particle.
- **Absorption**: When the incident light undergoes an interaction with the chromophore of a pigment or dye, light is absorbed and its energy is transferred into the absorbing molecule. This results in a reduction of intensity of radiation while passing through the medium. According to the *Lambert Beer* law, the extent of absorption is proportional to the concentration of the absorbing species.
- **Transmission**: Due to the combined effects of reflection, scattering and absorption the intensity of radiation is reduced while passing through a medium. This effect is quantified by measuring the ratio of the transmitted light to the incident light (I/I_0).

All these properties are responsible for a certain color of the pigmented film. The occurrence and extent of these effects depends on the characteristic of the medium, which are determined by the specific pigments used, the formulation components, as well as by the wavelength of the light.

In coating and ink applications, various pigment classes and pigment chemistries are used which influences the effects discussed before. Pigment classes that are currently used are divided into particle like (color) pigments and platelet like (effect) pigments. Particle like pigments can have all different types of absorption colors spanning from yellow to blue, as well as black and white. All these colored pigments have scattering, absorption, and transmission properties characterized by their chemical composition, particle size and particle shape. This class of pigments can be further divided into the following subgroups: colored organic pigments, white pigments, colored inorganic pigments, and black pigments.

Due to their geometry and the resulting optical effects, platelet like pigments shows interference colors in addition to absorption colors and different reflection and transmission effects. This class of pigments can be further divided into opaque platelet shape pigments (Figure 118) for example aluminum flakes, or semitransparent platelet shape pigments (Figure 119) such as interference pigments.

For the selection of a pigment for a UV curable coating it is not only important to look at its color in the visible range of the spectrum, but it is also important to assess its properties in the UV region. During the UV curing process the pigment and photoinitiator in fact compete for the incident light. Therefore, in order to allow the efficient curing of pigmented coatings, both the concentration and the optical properties (in the wavelength range between 200 and 420 nm) of the pigment are important. Ideally a pigment should have a so called "optical window" in the UV region where an absorption minimum of the pigment is overlapping with an absorption maximum of the photoinitiator. In addition to the absorption characteristics of the pigment in the UV region, the ability to scatter and/or reflect light at these wavelengths is also important. Scattering and reflection considerably increases the path lengths of the incident UV light within the film. According to the *Lambert Beer* law the absorbance by any molecule, is proportional to the path lengths of light in the medium, thereby also increasing the yield of initiating radicals formed by the photoinitiator. A pigment that is transparent in this wavelength domain does not compete with the photoinitiator for the incident light and hence is suitable for UV curing.

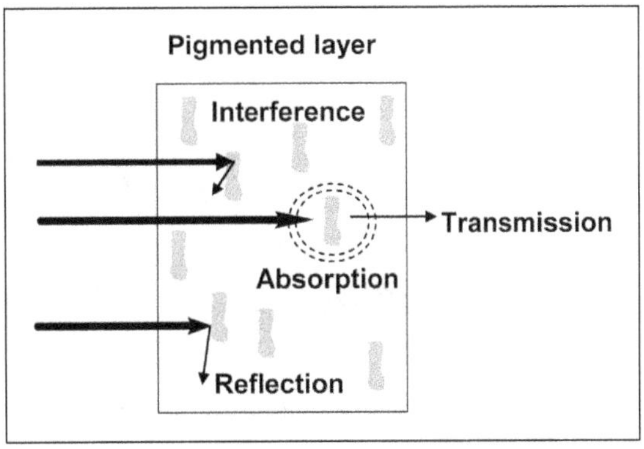

Figure 119: Optical effects of a semitransparent platelet shaped pigment

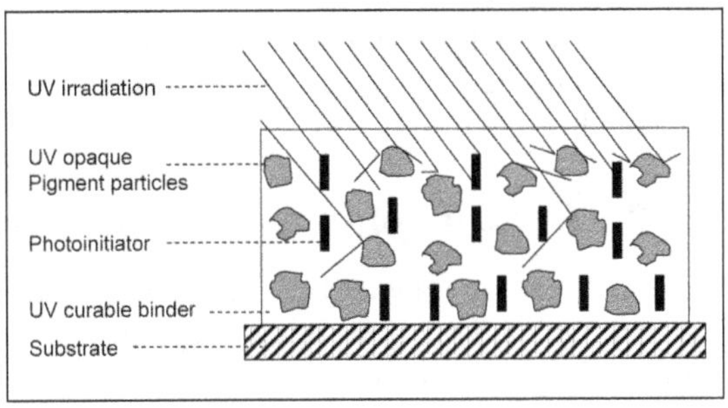

Figure 120: UV curing of a pigmented coating system with particle shaped pigments

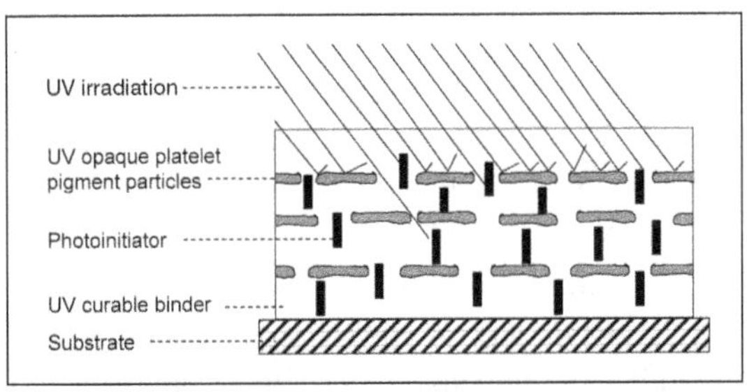

Figure 121: UV curing of a pigmented coating system with UV opaque platelet shaped pigments

Figure 120 shows that in a pigmented UV curable system the UV radiation do not reach the photoinitiator molecules in deeper layers by the direct route. In fact, molecules within the shadow of pigment particles can only be reached by scattered radiation. Due to the decreasing intensity of light passing through the formulation, the probability that a photoinitiator molecule can absorb UV radiation of a suitable wavelength decreases with increasing layer thickness. This is also true for opaque plateletshaped pigments such as metallic pigments (Figure 121). When the pigment absorbs all light, either due to a high concentration or to a thick coating resulting in long path lengths of light in the formulation, the photoinitiator is not able to absorb UV light anymore and therefore can not undergo a cleavage reaction producing initiating radicals. Therefore, the coating will remain uncured [124, 125].

The situation is different when, pearl luster pigments, which are transparent to UV radiation, are used (Figure 122). In this case, UV radiation can pass through the UV transparent platelets having a parallel orientation and penetrate down to the base of the coating even in the case of a thick coating. This also holds for other UV transparent particles such as inorganic fillers.

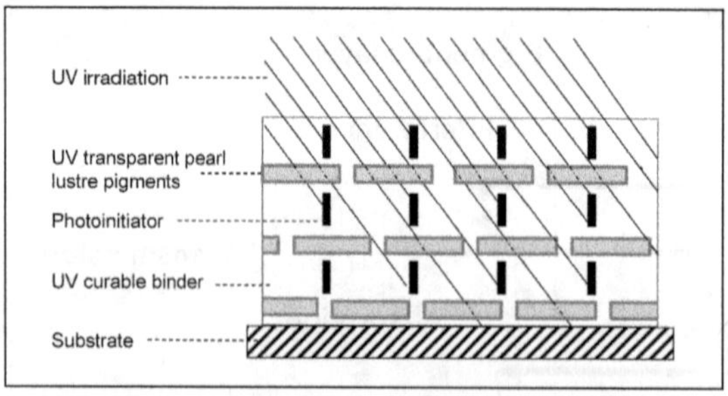

Figure 122: UV curing of a pigmented coating system with UV transparent platelet shaped pigments

The UV transmittance of a pigmented coatings film can be measured directly by UV spectroscopy, using the coating film preparations on UV transparent substrates as described in [126]. This technique involves an integration sphere detector, which is sensitive to transmitted radiation irrespective of the direction of the radiation. In this way all UV light capable of initiating the curing of the film is detected. This includes also light which is scattered by the pigment, and hence gives a better indication on the UV curing behavior of a pigment than a simple measurement of the transmission spectrum of the pigment. Even more parameters for the prediction of the curability of colored pigmented coatings have been included in a simulation model [127]. In this simulation, the transmission of the pigment, the absorption of the photoinitiator, the maximum film thickness and the type of the UV lamp are included.

3.6.2 Organic pigments in UV curable coatings

Organic pigments provide brilliant deep shades whereas inorganic pigments provide duller shades. This is mainly due to the stronger and steeper absorption/transmission characteristics of organic chromophores supported by the generally smaller particle size. Furthermore, organic pigments provide as single pigments or in mixtures generally more transparent coatings due to their smaller refractive indexes. An analysis of the absorption spectra of these pigments can explain some of the basic curing properties of the corresponding pigmented coatings.

Red has a high transmission in the range of 380 to 420 nm, which should principally facilitate the UV curing with suitable photoinitiators absorbing in this spectral range. However, a high concentration of these pigments is required in order to achieve good hiding or opacity. As a consequence the UV absorption is increased and the coatings are difficult to cure.

Blue pigments are characterized by high transmission in the relevant UV range and high hiding power. These conditions are favorable for the activation of the photoinitiator by UV light, and hence good curing properties can be achieved. Yellow pigments have a low transmission level in the UV and additionally these pigments have poor hiding power, requiring a high pigment level for an acceptable hiding. Due to these unfavorable conditions, through cure properties are usually poor. Green pigments have the lowest UV transmission level. Thus, green pigmented UV coatings with good hiding power are very difficult to cure by UV light.

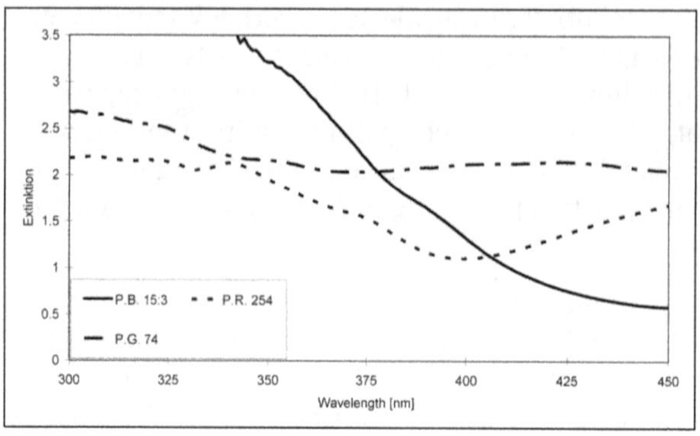

Figure 123: Absorption spectra of some colored pigments

It is difficult to forecast the cure properties when considering only the absorption spectrum (Figure 123) of a pigment, since the hiding power of the film has also to be considered. In fact, the hiding power of the pigment determines the

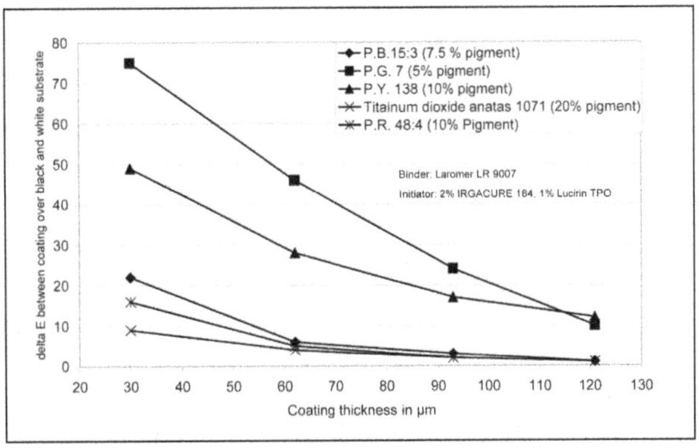

Figure 124: Hiding power of UV coatings - dependence on the film thickness

required pigment concentration, which in turn plays an important role in the curing properties. The maximum pigment concentration of a yellow, red, green and blue pigment has been determined at which a good through cure in full shade is still possible. The corresponding coatings - with the maximum pigment concentration - were applied on a black and white substrate and cured. Afterwards Δ E values between both areas were measured [128]. The results obtained confirm the practical experience: Green and yellow pigmented formulations are much more difficult to cure than blue pigmented UV coatings to give lacquers with satisfying hiding power (Figure 124) [128].

Note

In order to achieve good hiding power at the lowest pigment concentration the pigment has to be wetted and dispersed as well as possible (see Chapter 3.5.1).

As mentioned in Figure 124 blue colors with a satisfactory hiding power are most easy to cure while yellow and green are the most critical colors. In the coating industry, especially in wood coatings, mixtures of colored and white pigments are used, which are usually easier to cure by UV light than full-shade formulations. This is due to the fact that white TiO_2 pigments efficiently filter off the light below 380 nm, but are transparent at higher wavelengths where suitable photoinitiators of the acylphosphine oxide type (see Chapter 3.2.2.1) can efficiently absorb the light.

Different colors and pigment chemistries result in different color strengths. To make a fair comparison of the curability between the colors of a different strength, the ISD (international Standard Depth DIN 53235) is used. In order to determine the maximum curable wet film thickness and the correlating opacity of pigments, the pigments mentioned in the following paragraphs were tested in a polyester acrylate binder thinned with hexanediol diacrylate and trimethylol triacrylate. The single pigments were incorporated into the system at a concentration of 8.6 % for colored pigments and 18 % for titanium dioxide pigments by adding a suitable dispersant at a concentration of around 30 % of BET value, depending on the pigment used. From these pigmented formulations the white reduction 1/25 ISD according to DIN 53235 was prepared. A photoinitiator combination of 2.4 % 2-hydroxy-2-methyl-1-phenyl-propan-1-on (HMPP) and 0.6% BAPO was used to initiate the polymerization. Irradiation was carried out with two mercury medium pressure lamps of 80 W/cm lamp power each at a conveyor speed of 3 m/min.

In Figure 125 a selection of pigments is displayed, which are examples for each color space. The maximum curable wet film thickness was determined as the maximum slit coater thickness which still allowed the curing of the coating with full through cure. A higher thickness would prevent complete through cure. The formulations were applied on black and white panels in order to determine their opacity by measuring the Δ E value. Small Δ E values denote greater hiding.

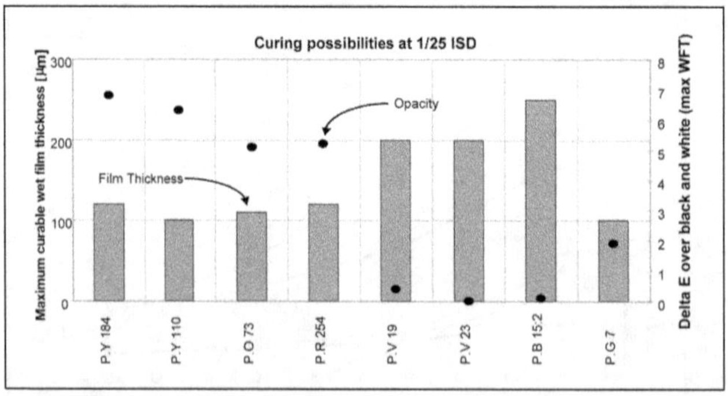

Figure 125: Overview of maximum curable wet film thickness and the correlating opacity of pigments representing different color spaces, ISD = standard color depth

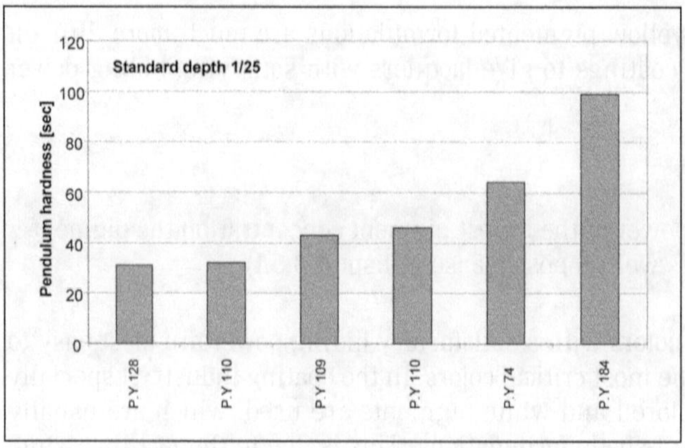

Figure 126: Through cure properties of different yellow pigments applied in a ISD 1/25 with a wet film thickness of 40 µm

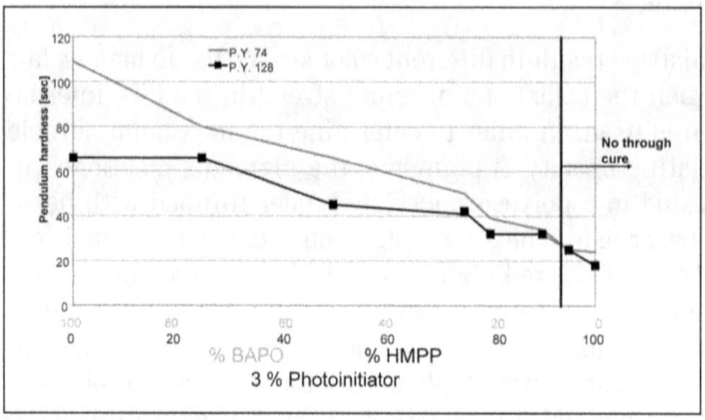

Figure 127: Cure properties of a yellow pigmented coating showing the dependence on the BAPO concentration

As shown in Figure 125, it is difficult to achieve full hiding with a yellow pigmented coating that allows good through cure properties. Approaches to optimize a yellow pigmented UV curable formulation for the best curing properties are shown in the next Figures. Figure 126 illustrates the through cure properties of different yellow pigments. For a comparison of the pigments, the 1/25 ISD formulations were applied at the same wet film thickness of 40 µm. After curing the pendulum hardness was determined. Under these conditions the inorganic bismuth vanadete P.Y. 184 shows the highest pendulum hardness. P.Y. 74 gives the highest hardness of all organic pigments.

The careful selection of the photoinitiator is another key factor for the successful UV cure of pigmented coatings. In the case of opaque and semi-opaque, formulations the use of a combination of two photoinitiators, one providing through cure and the other contributing mainly to surface cure, is recommended. BAPO, has an absorption band in the UVA which extends well into the visible light range. This photoinitiator is a key component for efficient through-cure of coatings. HMPP is an α-hydroxyketone photoinitiator and it absorbs at short UV wavelength and provides excellent surface cure. Figure 127 shows how the through cure properties of yellow pigmented coatings containing P.Y. 74 or P.Y. 128 depend on the ratio of BAPO to HMPP. The total photoinitiator concentration was at 3 %.

α-hydroxyketone photoinitiators (such as 1-[4-(2-hydroxyethoxy)-phenyl]-2-hydroxy-2-methyl-1-propan-1-on (HPMP)) or phenylglyoxylates, absorb only at shorter UV wavelengths and are mainly used to get a proper balance between surface and through cure. The contribution of different photoinitiators absorbing at short wavelengths to the through-cure of yellow pigmented coatings is shown in Figure 128 [129].

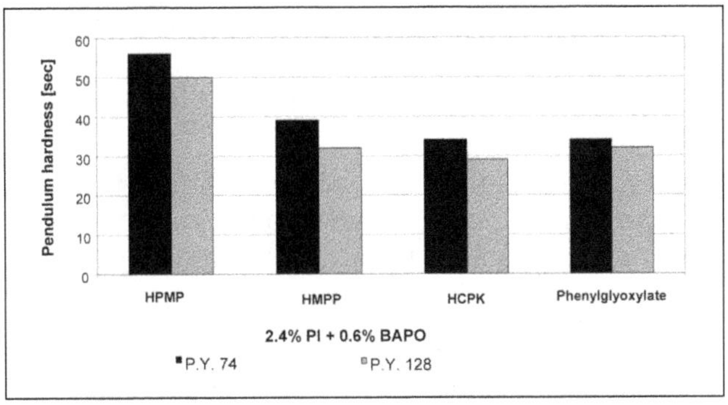

Figure 128: Contribution of different short wavelength absorbing photoinitiators to the through cure of pigmented coatings

Knowing the effect of the different yellow pigments on cure properties, three different versions of Maize Yellow (RAL 1006) and Safety Yellow were formulated (cf. Table 27). The maximum wet film thickness was determined as described above. The pendulum hardness of all coatings was measured at the lowest possible film thickness. The results show that shades composed of yellow pigments, which are more suitable for UV curing, are curable with higher film build than others (Figure 129).

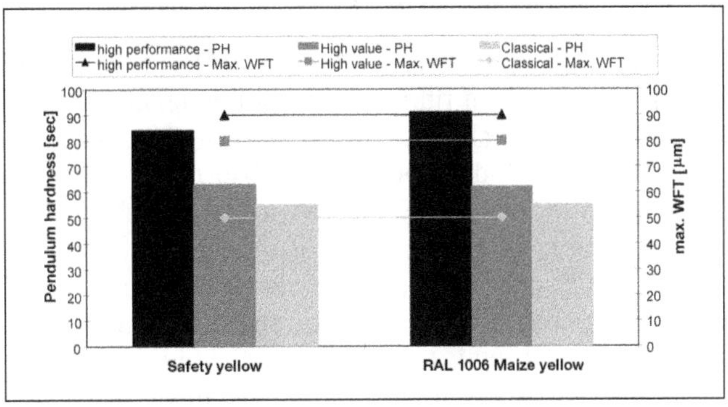

Figure 129: Formulation examples of Safety Yellow and Maize yellow (PH: pendulum hardness, max. WFT: maximum wet film thickness)

As described above it is possible to formulate UV curable coatings with full hiding power in the blue and magenta color space without a major limitation regarding the pigments used. It is much more difficult, however, to formulate in the green, red, and yellow color spaces. Despite all developmental efforts done over the past years on pigment development and new photoinitiators, limitations still exist. Approaches aimed at overcoming these limitations are:

Table 27: Composition of different yellow shades

Pigment recommendations for Safety Yellow

High performance		High value		Classical	
P.Y. 184	99.1	P.Y. 194	98.0	P.Y. 74	53.2
P.Y. 110	0.9	P.Y. 110	2.0	P.R.112	0.1
				titanium dioxide	46.7

Pigment recommendations for Maize Yellow

High performance		High value		Classical	
P.Y. 184	86.8	P.Y. 194	56.3	P.Y. 74	70.7
P.Y. 110	11.7	P.Y. 110	32.5	P.Y. 112	0.6
P.Y. 101	1.5	P.Y. 42	11.2	P.Y. 42	27.0
				titanium dioxide	1.7

1) the application of several layers on the top of each other,
2) the use of the dual cure technology, or
3) electron beam (EB) curing to achieve green, red, and yellow coatings with full hiding power.

In the case of dual cure the UV technology is used in combination with a second curing mechanism. An especially useful solution consists of dual cure systems where the light induces free radical polymerization combined with polyaddition reactions (see Chapter 3.4.3) [128].

By using EB curing, a relatively efficient polymerization of the binder system in presence of pigments can be achieved, providing an efficient through cure. The use of pigments in these systems is, however, limited due to color changes, which occur upon the exposure to EB radiation [130].

3.6.3 White pigments in UV curable coatings

White coatings can be achieved using different types of inorganic white pigments. Mostly used are magnesium titanate, zinc sulphide, and titanium dioxide in its anatase or rutile forms. Due to their chemical composition and particle size, these pigments provide different shades of white, hiding power and absorption characteristics, especially in the UV area. Magnesium titanate shows short wavelength absorption, which is most favorable for UV curing since it interferes least with the photoinitiator absorption. However, this pigment provides only poor performance characteristics regarding durability, hiding power and whitening power and is thus little used. Titanium dioxide in its rutile configuration has the most bathochromic absorption of all white pigments discussed. Rutile has thus the best performance in applications with regard to afore the mentioned properties. The absorption spectra of these inorganic white pigments are shown in Figure 130 [131].

Rutile-type titanium dioxide is usually preferred as the pigment in white coatings due to its high hiding power. It also acts as an efficient screen for the light of all wavelengths below 380 nm. A comparison of the absorption characteristic of the rutile pigment and the absorption spectra of the photoinitiators discussed in (see Chapter 3.2.2.13) reveals that only a few initiators are available on the market which are suitable to cure high performance white pigmented coatings efficiently.

A photoinitiator useful for white pigmented formulations must not impart yellowing to the cured coating. On the other hand, the fact that light below 380 nm is not available for the initiator implies that it must have an absorption extending into the visible range of the spectrum. This apparent contradiction can only be resolved when a photoinitiator is used that can absorb light in the long UV-A/visible range, but also undergoes complete bleaching of the absorption at the blue end of the visible spectrum, resulting in colorless products. The first class of photoinitiator which could fulfill this requirement is monoacylphosphine oxides [132]. More recently bisacylphosphine oxide photoinitiators were introduced, due to their extremely high absorption at 400 nm, and are

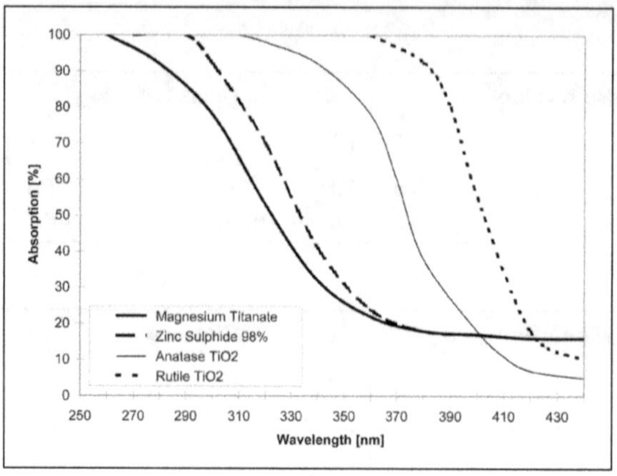

Figure 130: UV Absorption characteristic of different white pigments

especially suitable for applications such as highly pigmented white coatings [133]. In Figure 131 the absorption spectra of various photoinitiators are compared with the light cut off of rutile type titanium dioxide. The effect of these different photoinitiators on the pendulum hardness of a UV cured polyester acrylate coating containing a rutile titanium dioxide load of 25 % is shown in Figure 132.

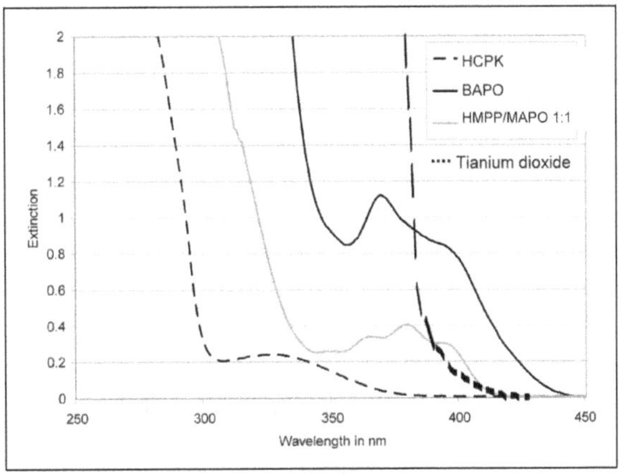

Figure 131: Absorption spectra of 0.05 % photoinitiators in acetonitrile and titanium dioxide rutile

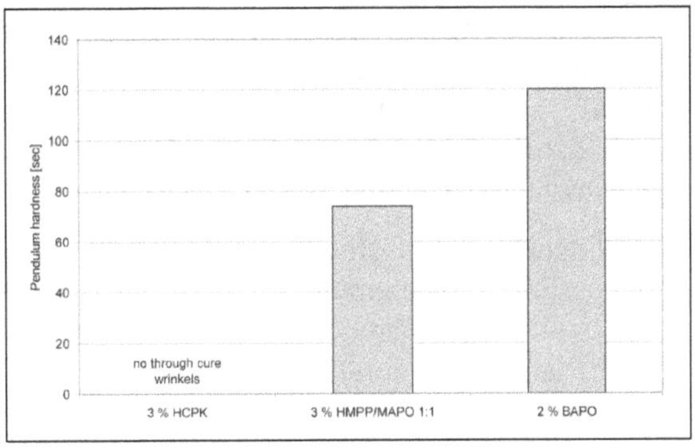

Figure 132: Curing results of a polyester acrylate coating containing 25 % titanium dioxide (rutile)

3.6.4 Black pigments in UV curable coatings

To create a black pigmented coating one can consider different classes of pigments. The possibilities are carbon blacks, iron oxide blacks, mixed metal oxide blacks and mixed phase pigments such as hematite structures. Most often used in the industry are carbon blacks and to a lesser extent iron oxide blacks. The focus of this chapter is on UV curing of black coatings using carbon black pigments.

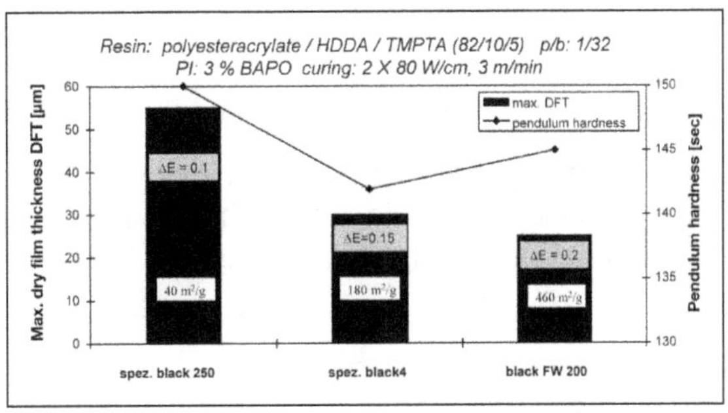

Figure 133: Cure properties of different carbon blacks in a UV curable polyester acrylate coating

Carbon black pigments have excellent application properties like excellent migration fastness, high light and weather durability, and a high deepness of color [134]. Carbon blacks absorb radiation both in the UV and visible light regions. Thus, they do not offer a well defined “spectral window” with reduced absorption that could be used for radiation curing. On the other hand, carbon black pigments give high color strength even at low pigment concentrations, and thus allow for a compromise between both sufficient opacity and sufficient through cure within certain limits of film thickness [135]. As for all pigments, the absorption behavior related to the particle size is also relevant for the different black pigments used for UV curing applications. Thus, certain carbon blacks like “Special Black” 250 with higher particle size normally intended for use in printing inks are frequently recommended for UV formulations. The effect of different particle sizes on the UV curing behavior is shown in Figure 133 [136].

It is important to note that the curing properties of opaque pigmented coatings also depend strongly on the formulation characteristics, irradiation conditions, and substrate reflection.

Nevertheless, it is the type of photoinitiators, pigments and extenders which affect the curing properties to a large extend. In the case of photoinitiators, bisacyl phosphine oxides have been reported to be particularly effective in black pigmented UV curable coatings [136-138].

In principle black pigmented coatings with satisfactory hiding power can be UV cured by using an optimized combination and selection of the type of carbon black, photoinitiator with a high quantum yield (such as a bisacyl phosphine oxide), and binder systems of high reactivity. Such black pigmented coatings are already well established in the industry, for example to paint three dimensional objects like motor blocks or wind screen wipers but also in the printing ink industry.

3.6.5 Effect pigments in UV curable coatings

Different effect pigments are on the market, which can be differentiated according to their transparency, chemical composition and the effect they are producing. Most often mica pigments are used. The coloristic feature of these pearlescent pigments is their lustrous and brilliant appearance. This is a consequence of the transparency and lamellar structure of the pigments, which consists of materials of different optical densities. Multiple reflections of incident light within this structure results in optical interferences, producing a chromatic effect which varies with the angle of observation. Color variation is achieved by coating on the mica particles different metal oxides with a specific layer thickness. Variations in the type of metal oxide and its film thickness provide a broad range of hue and intensity.

Mica pigments are classified as follows:

- TiO_2 - silver mica pigments (brilliant white) and interference mica pigments (dual color)
- Iron oxide - high luster earth tones (vibrant color with improved hiding power, especially red metallic like colors)
- Combination (TiO_2/Fe_2O_3 or TiO_2/Cr_2O_3) - color pigments (strong interference effect with improved hiding power due to additional mass tone color, especially gold and green mica pigments) [125].

Pearlescent mica pigments treated with TiO_2 provide a better UV transparency than opaque mica pigments treated with iron oxide. In this respect they are principally similar to titanium dioxide, zinc sulphide and magnesium titanate pigments. UV radiation can pass through the transparent platelets and thereby penetrate to the base of thick coatings.

Metal effect pigments, e.g., based on aluminum, do not have any UV transparency. They may, however, act as thousands of tiny mirrors in the coating. Hence the path lengths of the light is considerably increased by scattering, which increases the probability for light to hit a photoinitiator molecule and hence to start the polymerization. The preferred photoinitiator to be used in this case are ones that have a high quantum yield, like the bisacyl phosphine oxides.

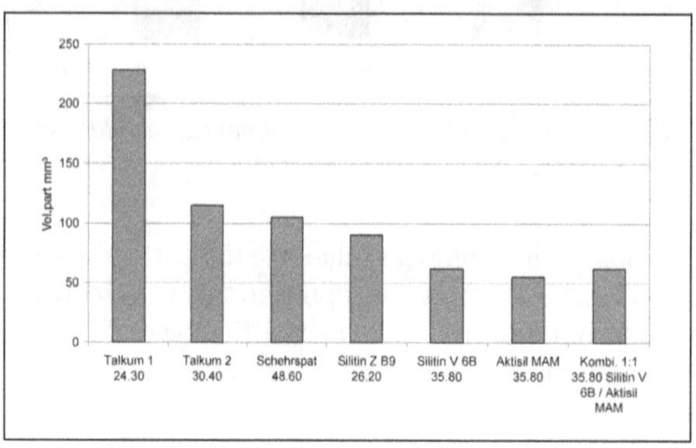

Figure 134: Abrasion resistance according to DIN 53 754 "Taber S 42" (0.5 Kg; 55 U/min; 100U)

3.6.6 Fillers

In addition to pigments, fillers are often used as extenders in UV curable coatings, to enhance coating properties. They affect mechanical, physical and optical properties, e.g., abrasion resistance, rheology, shrinkage, and matting (see Chapter 3.5.7). By proper selection of the structure and refractive index of the filler material, fully transparent coating films can be obtained [139]. Fillers are used mainly in UV curable wood coatings, e.g., parquet flooring. Figure 134 shows abrasion resistance properties according to DIN 53 754 of different fillers in an epoxy acrylate coating. Talc, barythes and siliceous earth were used to increase the abrasion resistance; siliceous earth shows the best results [140].

References

[1] Tego Journal, EVONIK Tego Chemie Service GmbH, 3rd Edition, Essen, Germany, 2007
[2] T. Brock, M. Groteklaes, P. Mischke, Hrsg. U. Zorll, Lehrbuch der Lacktechnologie, 2. berichtigte Auflage, Vincentz Verlag, Hannover, 2000
[3] C. Decker, Materials Science and Technology, Processing of Polymers, ed. by H.E.H. Meijer, VCH Verlagsgesellschaft mbH, Weinheim, 18 (13), 1997
[4] J.V. Crivello, K. Dietliker, ***Chemistry and Technology of UV&EB Formulation for Coatings, Inks & Paints***, G. Bradley Ed., John Wiley and Sons, New York, Sita Technology Ltd, London, Vol. 3, 1998
[5] K. Dietliker, A compilation of photoinitiators commercially available for UV today, SITA Technology, Edinburgh & London, 2002
[6] W. Rutsch, K. Dietliker, D. Leppard, M. Koehler, L. Misev, U. Kolczak, G. Rist, Prog. Org. Coat., 27, 1996, p. 227
[7] G. Berner, R. Kirchmayr, G. Rist, Technical Conference Procedings - Int. Conf. Org. Coat. Sci. Technol., 3, 1977, p. 204
[8] K. Studer, R. Koeniger, Eur. Coat. J., 1-2, 2001, p. 26
[9] R. Burrows, R.S. Davidson, D.R. Illsley, Surf. Coat. Int. Part B, 87 (B2), 2004, p. 127
[10] L. Felder, R. Kirchmayr, R. Huesler, Eur. Pat. Appl. EP 3002 19790711 (1979)
[11] R. Kirchmayr, G. Berner, R. Huesler, G. Rist, Farbe + Lack, 88, 1982, p. 910
[12] C. Decker, K. Moussa, Makromol. Chem., 189, 1988, p. 2381
[13] K. Studer, C. Decker, E. Beck, R. Schwalm, Prog. Org. Coat., 48, 2003, p. 92 and 101
[14] W. Rutsch, G. Berner, R. Kirchmayr, R. Huesler, G. Rist, N. Buehler, Organic Coatings, 8, 1986, p. 175
[15] R. S. Davidson, Radiation Curing in Polymer Science and Technology, Volume III, J.P. Fouassier and J.F. Rabek editors, Elsevier Applied Science, London, 1993, p. 153
[16] I. Gatlik, P. Rzadek, G. Gescheidt, G. Rist, B. Hellrung, J. Wirz, K. Dietliker, G. Hug, M. Kunz, J.-P. Wolf, J. Am. Chem Soc., 121, 1999, p. 8332
[17] C. Decker, K. Zahouily, D. Decker, T. Nguyen, T. Viet, Polymer, 42, 2001, p. 7551
[18] R. Jahn, T. Jung, Prog. Org. Coat., 43, 2001, p. 50
[19] S. Hu, X. Wu, D. C. Neckers, Macromolecules, 33, 2000, p. 4030
[20] S. Hu, D.C. Neckers, J . Photochem. Photobiol., A, 118, 1998, p. 75
[21] W.D. Cook, Polymer, 33, 1992, p. 600
[22] J. Segurola, N. Allen, M. Edge, A. Parrondo, I. Roberts, J . Photochem. Photobiol., A, 122, 1999, p. 115
[23] J.V. Crivello, Ann. Rev. Mater. Sci., 13, 1983, p. 173
[24] J.V. Crivello, J. Polym. Sci., Part A: Polym. Chem., 37, 1999, p. 4241
[25] H. Ito, C.G. Willson, Polym. Eng. Sci., 23, 1983, p. 1012
[26] J.M. Frechet, Pure&Appl:Chem., 64, 1992, p. 1239
[27] S.I. Schlesinger, Polym. Eng. Sci., 14, 1974, p. 513
[28] C. Decker, Prog. Polym. Sci., 21, 1996, p. 593
[29] A. Fuchs, R. Huesler, C. Schregenberger, M. Kunz, PCT Int. Appl. WO2003040076 (2003)
[30] J.-P. Wolf, R. Huesler, W. Peter, R. Sommerlade, S. Boulmaaz, PCT Int. Appl. WO2003091287 (2003)
[31] W. Peter, S. Megert, D. Rogez, E. Sitzmann, Eur. Coat. J., 11, 2002, p. 14
[32] J. Rogers, G. Baudin, T. Jung, R. Oehrlein, G. Baisch, PCT Int. Appl. WO2004108799 (2004)
[33] G. Norcini, A. Casiraghi, M. Visconti, G. Li Bassi, PCT Int. Appl. WO2003008404 (2003)

[34] F. Cazaux, X. Coqueret, B. Lignot, C. Loucheux, P.J. Rousseau, Coatings Technology, Vol. 66 (838) 1994, p. 27 - 34
[35] R. Bongiovanni, G. Malucelli, M. Sangermano, A. Priola, Macromol. Symp., 187, 2002, p. 481 - 492
[36] D. James, P. Appelkvist, E. Gustavsson, RadTech Europe 2005, Conference Proceedings Vol. I, 2005, p. 553 - 559
[37] M. S.Chisholm, H.C. Harte, I. Rumoroso, A.T. Slark, M.C. Williams, Surface Coatings Int. Part A, 06, 2003, p. 221 - 228
[38] H.-G. Elias, Makromoleküle, Band 1, 5. Auflage, Hüthig & Wepf Verlag, Basel, Heidelberg, New York, 1990
[39] G. Odian, Principles of Polymerization, 3rd Edition, Wiley Interscience, New York, 1991
[40] B.A. Wolf, Chem. Phys. Macromol., 1991, p. 273 - 294
[41] J. Oliver, M. Bernard, H. Cavalie, RadTech Europe 2005, Conference Proceedings Vol. II, 2005, p. 199 - 206
[42] R. Frings, M. Wend, DIC Technical Review No. 9, 2003, p. 43 - 51
[43] B.A. Wolf, Pure and Appl. Chem., 57 (2), 1985, p. 323 - 336
[44] C. Decker, K. Moussa, Makromol. Chem., 189, 1988, p. 2381
[45] P. Garratt, Hrsg. Ulrich Zorll, Strahlenhärtung, Vincentz Verlag, Hannover, Germany, 1996
[46] E. Andrzejewska, M. Janaszczyk, M. Andrzejewski, RadTech Europe 2005, Conference Proceedings Vol. II, 2005, p. 49 - 56
[47] wwwsartomer.com/home.asp: Glass Transition Temperatures of Sartomer Products
[48] A. Hartwig, M. Popp, K. Teczyk, RadTech Europe 2005, Conference Proceedings Vol. I, 2005, p. 607 - 614
[49] BGIA-Report 1/05: Gefahrstoffliste 2005, Gefahrstoffe am Arbeitsplatz, Hauptverband der gewerblichen Berufsgenossenschaften
[50] wwwsartomer.com/home.asp: Sartomer: Shrinkage of UV Monomers
[51] A. Fukazawa, Ink World Magazine, June 2006
[52] A. J. Bean, Ink World Magazine, October 2002
[53] M. Osterhold, K. Armbruster, Macromol. Symp., 126, 1997, p. 295 - 306
[54] S. Smeets, E. Boerrigter, S. Peeters, ECJ, 2006, p. 42 - 48
[55] A. Carroy, L. Druene, M. Garinet, European Coatings Conference „Modern Coatings for Plastics Substrates II" 17/18 November 2005, Berlin, Germany
[56] G. Meichsner, M. Maday, W. Paulus, R. Schwalm, E. Beck, K. Menzel, Polymers & Coatings 2006, 24. - 26. September 2006, p. 29
[57] D. Stoye, W. Freitag, Resins for Coatings, Hanser Verlag, Munich, Germany, 1996
[58] P. Glöckner, H. Ritter, M. Osterhold, M. Buhk, W. Schlesing, Die Angewandte Makromolekulare Chemie, 269, Nr. 4679, 1999, p. 71 - 77
[59] R. Dowbenko, C. Friedlander, G. Gruber, P. Prucnal, M. Wismer, Progress in Organic Coatings, 11 (1983), p. 71 - 103
[60] R. Klagge, RadTech Europe 2003, Conference Proceedings Vol. I, 2003, p. 391 - 399
[61] T. Fäcke, W. Fischer, D. Rappen, F. Richter, J. Weikard, RadTech Europe 2003, Conference Proceedings Vol. I, 2003, p. 217 - 224
[62] D. Hoppe, R. Lomölder, E. Spyrou, F. Plogmann, A. Raukamp, P. Speier, PU Magazin, 03, 01, 2003, p. 41 - 43
[63] M. Ludewig, N. Stöckel, J. Weikard, RadTech Europe 2005, Conference Proceedings Vol. I, 2005, p. 261 - 268
[64] L. Dulany, RadTech USA 2000, Conference Proceedings, 2000, p. 103 - 112
[65] R.S. Davidson, in J.P. Fouassier, J.F. Rabek, Radiation Curing in Polymer Science and Technology - Vol. III, Elsevier Applied Science, London, New York, 1993
[66] J. Oliver, RadTech Europe 2007, Conference Proceedings
[67] S. Herrwerth, RadTech Europe 2005, Conference Proceedings, Vol. I, 2005, p. 583 - 596
[68] S. Oestreich, RadTech Europe 2001, Conference Proceedings, 2001, p. 397 - 403
[69] M. L.Gould, S. Narayan-Sarathy, T.E. Hammond, R.B. Rechter, RadTech Europe 2005, Conference Proceedings Vol. I, 2005, p. 245 - 251
[70] H. Esselbrugge, Congress Proceedings, Nuremberg, Germany 2007, Paper VI.5
[71] H. Esselbrugge, European Coatings Journal, 2007, 6, p. 60 - 63

[72] A. Jacobine, in J.P. Fouassier, J.F. Rabek, Radiation Curing in Polymer Science and Technology - Vol. III, Elsevier Applied Science, London, New York, 1993
[73] C. Hoyle, T. Y. Lee, T. M. Roper, J. Polym. Sci., Polym. Chem. Ed., 2004, 42, p. 5301 - 5338
[74] www.coatings-colorants.com: Adhesion and Synthetic Resins
[75] www.roehm.com
[76] M. Heylen, Congress Proceedings, Nuremberg, Germany 2007, Paper IX.1
[77] M. Fies, H. Endres, R. Klagge, European Coating Show, Nuremberg, Germany, 2003
[78] www.specialchem4coatings.com: 07/2005: Radiation curing update and outlook
[79] W. Carter, M. Jupina, RadTech Europe 2003, Conference Proceedings Vol. I, 2003, p. 241 - 248
[80] www.chemicals.ucb-group.com: Uvacure cationic cure technology, Formulating guidelines
[81] E. Spyrou, Progress in Organic Coatings, 43, 2001, p. 25 - 31
[82] C. Decker, RadTech Europe 2001, Conference Proceedings, 2001, p. 213 - 218
[83] C. Decker, I. Lorinczova, RadTech Europe 2003, Conference Proceedings Vol. II, 2003, p. 669 - 677
[84] E. Spyrou, RadTech Europe 2005, Conference Proceedings Vol. II, 2005, p. 267 - 276
[85] C. Irle, M. Bayona, Farbe und Lack, 112, 2006, p. 26 - 29
[86] W. Fischer, E. Lühmann, D. Rappen, J. Weikard, European Coating Show, Nuremberg, Germany, 2001
[87] J. Weikard, W. Fischer, RadTech Europe 2001, Conference Proceedings, 2001, p. 431 - 436
[88] S. Sommer, H. Blum, R. Kopp, C. Irle, Congress Proceedings, Nuremberg, Germany 2007, Paper XVI.5
[89] P. Glöckner, 71. FH-Texte Januar 2005, Fachtagung an der Aachen University of Applied Sciences, Wässrige Kunststoffdispersionen und ihre Anwendung, 2005, p. 9 - 20
[90] M. Heylen, RadTech Europe 2007, Conference Proceedings
[91] J. Weikard, 4. Esslinger Oberflächenseminar „UV-Härtung von Oberflächen“, 20. und 21. September 2004
[92] E. Beck, ECJ, 04, 2006, p. 32 - 39
[93] S. Struck, “Flüssiger Pigment-Manager”, Farbe und Lack 06/2006, p. 35
[94] Tego Journal, Evonik Tego Chemie GmbH, 3rd Edition, 2007, Technical Background Pigment Wetting and Dispersing Additives, p. 78
[95] Tego Journal, Evonik Tego Chemie GmbH, 3rd Edition, 2007, Technical Background Pigment Wetting and Dispersing Additives, p. 84
[96] P. Cavaleiro, “Tailoring the side chains”, ECJ 04/2007, p. 120
[97] S. Murray, Elementis, New Additive Chemistry for Improved Pigment Suspension in Non-Aqueous UV Cured High Performance Coatings, ‚RadTech Congress Barcenona, 2005
[98] D. G. Miller, Cabot Corp., Rheology Control of High Solids Coatings
[99] Degussa Bulletin Fine Particles
[100] Tego Journal, Evonik Tego Chemie GmbH, 3rd Edition, 2007, Technical Background Substrate Wetting Additives, p. 72
[101] H. Semmler, Tego Chemie Service GmbH, Effective Substrate Wetting on Difficult-to-Wet Substrates, 7th Nürnberg Congress, 2003
[102] DuPont Zonyl Fluorosurfactants
[103] Tego Journal, Evonik Tego Chemie GmbH, 3rd Edition, 2007, Technical Background Deaerators, p. 55
[104] H. Semmler, Tego Chemie Service GmbH, Tailor-made additives for the elimination and prevention of microfoam in waterbased coatings, Paper, June 2001
[105] Tego Journal, Evonik Tego Chemie GmbH, 3rd Edition, 2007, Technical Background Defoamers, p. 48
[106] Tego Journal, Evonik Tego Chemie GmbH, 3rd Edition, 2007, Technical Background Deaerators, p. 57
[107] E. Reuter, Moderne Additive für strahlenhärtende Beschichtungen, Farbe und Lack, 08/2000, p. 24
[108] H. Becker, H. Vogel, Chem. Eng. Technol, 17, 2004, p. 1122
[109] L. Levy, Plant/Operations Progress, 6(4), 1987, p. 188
[110] S. Villeneuve, E. Pighetti, S. Ilg, American Inkmaker, May 2004, p. 18
[111] E. Krendlinger, Clariant GmbH, Waxes for UV and EB curing coatings, Radtech Conference

Basel, 2001
[112] M. Biehler, BASF AG, Influence of diefferent parameters on the scratch resistance of UV coatings, Radtech Conference Basel, 2001
[113] V. Schrenk, Evonik Tego, Presentation: Vestowax – Products and Applications, 2007
[114] Tego Journal, Evonik Tego Chemie GmbH, 3rd Edition, 2007, Technical Background Slip, Flow and Radiation-Curing Additives, p. 63
[115] S. Struck, Radiation-Curable Silicone-Based Additives for Coatings and Printing Inks, publication, www.tego.de
[116] Degussa Bulletin Fine Particles
[117] Tego Journal, Evonik Tego Chemie GmbH, 3rd Edition, 2007, Technical Background Slip, Flow and Radiation-Curing Additives, p. 63
[118] H.-D. Christian, EVONIK Degussa, "New Methods for Matting of 100 % UV Coatings", Rad Tech Europe 2005, p. 279
[119] P. Kelly, Akzo Nobel UV Resins, Rheological Parameters of UV/EB Curing Matt Systems Radtech Conference Basel, 2001
[120] Degussa 4 radiation-curing Coatings & Inks
[121] Johnson Matthey, Catalysts Brochure, 2003
[122] R. Hume Duncan, US Patent 5821276, 1998
[123] Nordic Paint association congress 2003, New investigations in Pigmented UV-coatings, M. Biehler, T. Frey, K. Graf, K. Menzel, P. Enekel, H.-H Bankowsky, E. Beck
[124] K. Dorfner; UV Photoinitiators in pigmented systems, Radiation Curing of Polymers II; Royal society of Chemistry; 1990; No 89; P 216ff
[125] R. Maisch, O. Stahlecker, M. Kieser ECJ 9/1994; p 582-601
[126] R. Jahn, T. Jung, 22nd Int. Conf. Org. Coatings, Athens, July 2000
[127] T. Frey, M.Biehler, K. Graf, RadTech Europe 2005 Conference proceedings p. 515
[128] M. Biehler, T. Frey, K. Graf, K. Menzel, P. Enekel, H.-H. Bankowsky, E. Beck; Nordic Paint association Congress; 2003; proceedings p. 87ff
[129] R. Genuneith Diplomawork „Effective UV-Curing of colored Pigmented Systems" July 2002
[130] R L. Dulog, H. Schweiger, Farbe & Lack 103.Jg, 8/97, p. 30.
[131] P. G. Garratt, Proceeding 3rd Int. Symp. Manchester No 89 Sept. 1990, p.103
[132] M. Jakobi, A. Henne, A.Boettcher, Polm.Paint Col. J. 1985, 175, p. 636
[133] H. F. Gruber, Prog. Polym. Sci., Vol 47, 1992, p. 953
[134] Ullmann18, Band 18 p.593
[135] N. Pietschmann Farbe & Lack 100 (1994), no 11, pp 923-929
[136] T. Jung, Farbe & Lack 104 (1998), no 12,pp 81-85
[137] J. Segurola, N. S. Allen, M. Edge, A. Parrondo and I. Roberts, J. Coatings Technol. 71 (1999), no 894, pp 61-67
[138] N. Pietschmann, I. Peter, P. Stroh, European Coatings Journal (2000), p. 7
[139] Ullmann Encylopedia band 18 p. 647
[140] S. Heckel, H. Oggermüller Farbe & Lack, Jahrgang 103, 10/97, p. 70

4 Radiation sources

In radiation curing usually photoinitiators and binders are considered to be the most important elements of this technology. This is true for the chemistry. Before starting the development of a new coating or ink formulation the physical part of the radiation technology used should be considered carefully in order to adjust the formulation components to the parameters of equipment used to achieve best cure results. Most of us are familiar with the visible light spectrum, which is part of the more extended electromagnetic spectrum (Figure 135). The full spectrum of light (including infrared, visible, ultraviolet and X-ray) is differentiated by the wavelength, which is specified in nanometer (nm). In radiation curing two major types of radiation are used electron beam (EB) and UV/Vis light. While the latter is electromagnetic radiation of different energy (mostly in the wavelengths range of 200 to 450 nm) for initiating the radical polymerization of oligomers, EB uses highly accelerated electrons with energy of 100 to 250 eV.

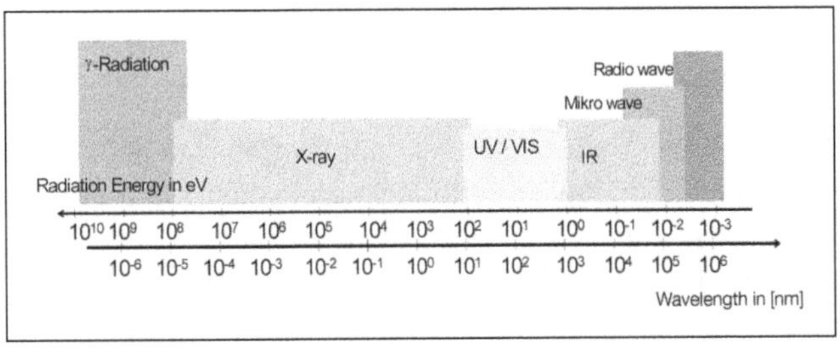

Figure 135: Electromagnetic spectrum

4.1 UV radiation technology

Even though the human eye cannot see UV light, this portion of the spectrum is part of sunlight and plays an important role in radiation curing. According to DIN 5031, UV light is further split down into long wavelength UVA light (315 to 380 nm), medium wavelength UVB light (280 to 315 nm) and short wavelength UVC light (100 to 280 nm). Different photoinitiators are currently available that respond to different UV wavelengths of light when irradiated.

To comply with regulatory requirements and due to several new applications raw materials as well as the physical equipments have been optimized. For example, over the last 10 years there has been a dramatic increase in the number of photoinitiators and the types of lamp technologies. Historically mercury medium pressure lamps are still the most common lamps used. More recent developments use UVA light lamps, fluorescent lamps, LED´s or plasma as UV radiation sources. Only a fraction of the energy applied to a lamp is transformed into an output in the form of UV radiation. For curing of films it is very important to know how much UV light reaches the substrate. To measure that, the UV energy (dose) and the UV irradiance (intensity) can be determined either by simple measuring strips or by more complex photometers.

UV energy (dose) [J/cm²]

This is the total UV energy deposed on the ink or coating surface and is inverse proportional to speed of the conveyor belt in the curing equipment. It is related to the total amount of energy available from the lamps and the time of exposure.

Glöckner/Jung/Struck/Studer: Radiation Curing for Coatings and Printing Inks

ISBN: 978-3-86630-907-4

Intensity of UV irradiance

This is the intensity of the light at the ink or coating surface and is characteristic of the lamp and geometry of the reflector. The intensity of the lamp system is independent on the irradiation time.

It is important to clearly distinguish between the UV dose and the UV irradiance [W/cm^2]. Both parameters influence the outcome of a curing process albeit at a different level. While a minimum UV dose is usually required to fully cure a formulation, the irradiance will govern the outcome of the curing process to a large extent. As an example, if a sufficient UV dose is applied to a radically curing formulation with a very low UV irradiance; insufficient cure may be obtained due to oxygen inhibition.

4.1.1 Light sources

4.1.1.1 Mercury medium pressure lamps

Mercury medium pressure lamps are used mostly for clear coat applications such as overprint varnishes and wood coatings. They usually have a bulb form and consist of a quartz tube containing small amounts of mercury. The tube is also filled with an inert gas like argon, which is essential to start up the lamp. In UV curing technology the most commonly used type is the mercury medium pressure lamp. Though there are differences in the way bulbs are powered, the general operation is the same. Energy is applied to the mercury, which causes it to heat and quickly vaporize. When a bulb reaches its operating temperature of approximately 1,000 °C, the vapor becomes a plasma and emits the characteristic emission spectrum of mercury. This emission consists of several distinctive lines in the UV and the visible range (Figure 136). The inert gas filling supports the transfer of mercury from a liquid to a plasma state.

The radiation power of such lamps is described in power/length (Watt/cm). The output of a single lamp is expressed in Watts. For industrial applications lamps of a length varying from 20 cm to 2 m are typically used. For special applications like PVC flooring, lamps with lengths of up to 5 m are used.

The electric power which is consumed by the lamp is converted by 20 to 25% into UV light, while 60 % of the energy is converted into IR radiation and the rest emitted as visible light. The IR radiation may heat up the substrate and this can often have a positive impact on curing. For example, heating reduces the viscosity of the formulation, thereby increasing the mobility of acrylate functions and hence the reactivity of the formulation (see Chapter 3.3.1). In cases where temperature sensitive substrates like films are used, the emitted IR heat needs to be reduced dramatically in order to prevent the damage of the substrate. This can be achieved by cooling as well as by the use of reflectors (which are transparent for IR radiation and therefore eliminate a large part of the IR radiation).

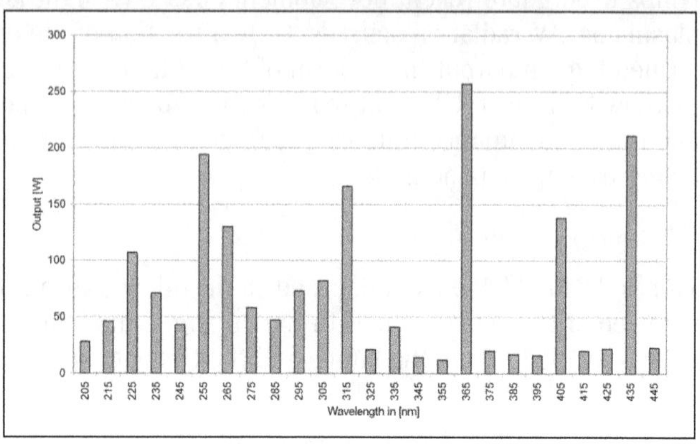

Figure 136: Spectral radiance of a mercury medium pressure lamp (Fusion UV-systems)

4.1.1.2 Doped mercury medium pressure lamps

For a better fit to the spectral window available in applications like pigmented coatings or UV curing in presence of UV absorbers, lamps with a more intense emission at longer wavelengths have been requested by the market in order to achieve better through cure. This shift to longer wavelength can be realized by the doping of the mercury with iron or gallium. Lamps containing iron doped mercury as a light generation medium show an increased UV output in the UVB and UVA light region (Figure 137). An increased output in the UVA region is obtained with a gallium doped mercury lamp (Figure 138).

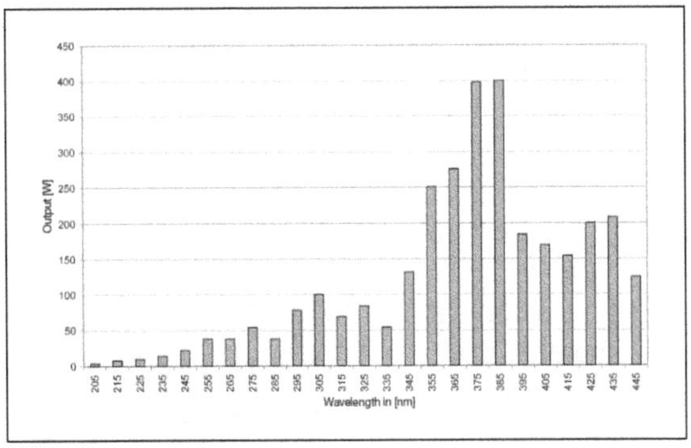

Figure 137: Spectral radiance of Iron doped mercury medium pressure lamp (Fusion UV-systems)

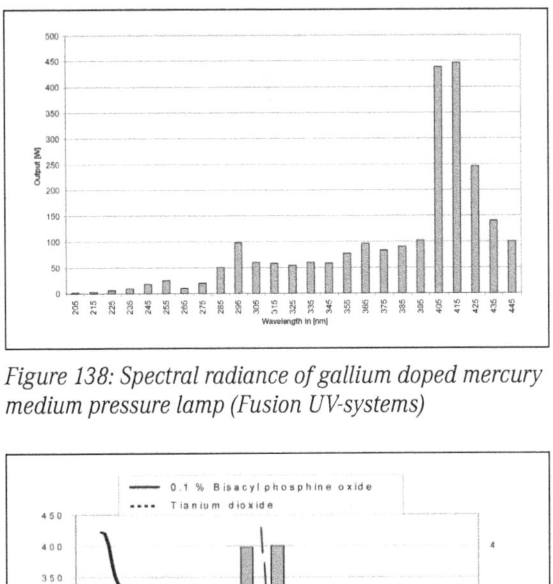

Figure 138: Spectral radiance of gallium doped mercury medium pressure lamp (Fusion UV-systems)

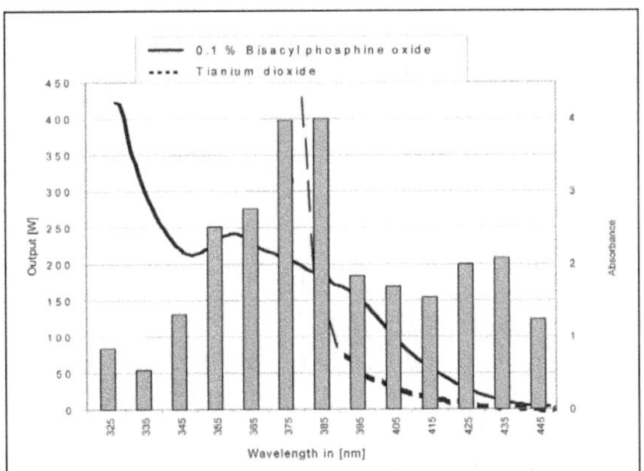

Figure 139: Spectral overlap of photoinitiator (bisacyl phosphine oxide), pigment (titanium dioxide) and lamp (iron doped mercury medium pressure lamp)

In case of a gallium lamp it is important to notice that this lamp has only a very low output in the UVB and UVC wavelengths. This can result in a lack of surface cure, an effect that is overcome in the industry by using a combination of gallium doped lamps and lamps filled only with mercury. In the curing unit the gallium lamp, responsible for an effective through cure of the coating, is used in a first place, followed by a mercury lamp responsible for the surface cure.

In order to obtain sufficiently crosslinked films the lamp emission, the absorption of the photoinitiator and of the pigments has to be taken into account. As discussed in Chapter 3.6 it is quite difficult to cure white pigmented films. An optimum match of transmission window of the formulation is crucial for an efficient curing. For example, this is the case for a white pigmented formulation cured with a bisacyl phosphine oxides photoinitiator and irradiated by an iron or gallium doped lamp. The absorption of the photoinitiator extends well above the cut-off edge of the titanium dioxide pigment into a wavelength range in which the iron doped lamp has a strong emission. This combination thus allows the efficient through-cure of a white pigmented lacquer (Figure 139).

4.1.1.3 Mercury low pressure lamps

The design of mercury low pressure lamps is similar to that of fluorescent lamps usually used to illuminate rooms. Due to the use of luminescent materials, the emission spectra can be shifted towards longer wavelengths. The spectral output can be designed by a large variety of phosphorus materials, to optimize the UVA spectra. Most of these lamps have a strong emission in the UVA range and are hence efficient for through cure.

TL 03 and TL 05 lamps are two typical representatives. The spectral output of a TL 03 lamp is shown in Figure 140. These lamps have a low power and work at low temperature. The relatively low light intensity (UV irradiance) of these lamps results in considerable oxygen inhibition which strongly influences the surface cure of radically curing coatings. The temperature increase on the substrate is negligible, since the convection heat of ca. 35 to 40 °C generated by these lamps is mostly eliminated through the air cooling system. In the industry such fluorescent lamps have been promoted for refinish coatings [1] (Figure 141) as well as for the release of base for photolatent precursors, in presence of photosensitizers (see Chapter 3.2.5).

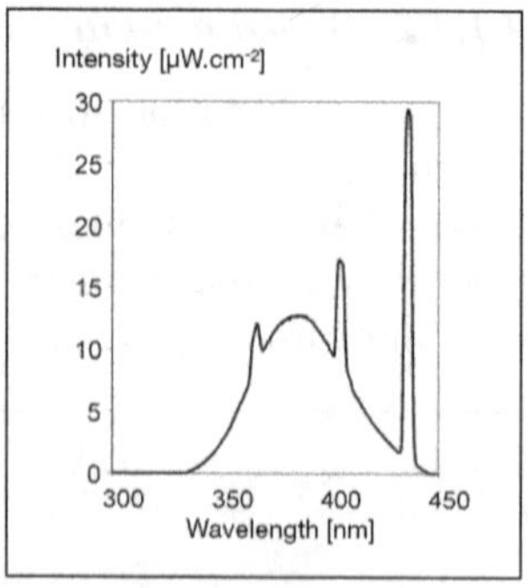

Figure 140: Spectral radiance of TL 03 and TL 05 lamps (Philips Lightning)

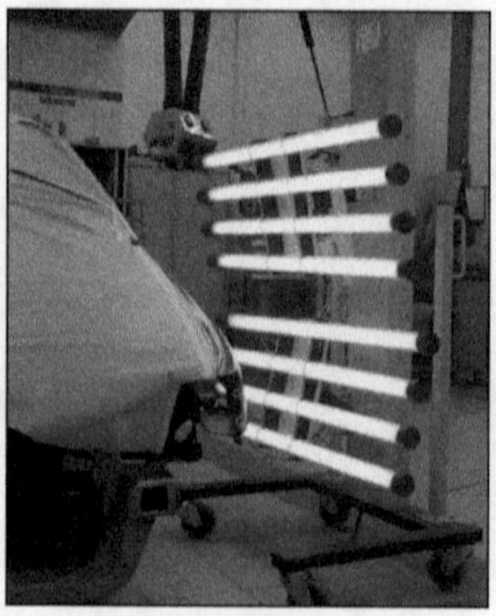

Figure 141: Drying of a 2K UVA curable coating [1]

4.1.1.4 LED arrays

Solid state light emitting diode (LED) arrays are currently proposed as novel light sources for UV curing technology. The most striking difference between solid state UV LED arrays and mercury lamps is the narrow spectral output that characterizes UV LED arrays. The output is a narrow peak with a bandwidth in the range of 20 nm. Currently, the LED arrays most useful for industrial coatings operate in the 390 to 410 nm regions with a maximum at 405 nm, due to the available LED technology and the cost/power trade offs. LED arrays emitting at shorter wavelengths are available (e.g., 365 nm) but currently provide a rather low light intensity and are more expensive. However, this new technology is experiencing rapid development aimed at more efficient devices emitting at lower wavelengths and increasing intensity.

UV LED arrays are constructed of nearly microscopic diode emitters arranged in solid state matrix. An advantage of this set-up compared to bulb-type lamps is the fact that these matrices are not inherently limited in shape or size [2, 3]. Thus the geometrical design of emitters that fits well to the object to be cured is possible.

4.1.1.5 Plasma curing

Besides the liquid, solid and gaseous states, plasma is mentioned as the fourth physical state. Plasma is usually generated by the ionization of gases by means of electrical energy. Plasmas are conductive but are electrically quasi neutral, because they have equal concentrations of positively and negatively charged carriers. In addition to the electrically charged particles (electrons, singly and multiply charged ions), plasmas may contain neutral molecules, atoms, and radicals. The electrical discharge is generally accompanied by intense emission of electromagnetic radiation which can be produced in the form of UV light with wavelengths between 200 and 380 nm, which can be used to activate photoinitiators and hence initiate the radical polymerization. Such UV radiation is produced for example in a microwave induced plasmas at low pressure (Figure 142).

The properties of the plasma can be adjusted to suit the requirements of a specific application by optimizing a number of process parameters, including gas type, gas pressure, current, voltage, external electric and magnetic fields, and electrode shape and material [4-6].

4.1.2 UV curing equipment

The UV lamps discussed above are usually integrated into a highly engineered UV curing equipment line consisting of an electrical power supply, cooling equipment for the lamps, reflectors, and an exhaustion device and in most cases a transport band. These features are combined to fulfill the requirements set by the application. In most cases the equipment is designed for processing two dimensional substrates transported piece by piece on a conveyor belt (Figure 143). Examples are curing units for the production of parquet flooring or role-to-role coatings. Such kind of UV equipment is usually a tailor-made installation designed to fulfill on the one hand the application's needs for an cure of UV curable ink or coating efficiently, and on the other hand to fit into the whole production process [7].

4.1.2.1 Power systems of mercury lamps

Mercury medium pressure lamps can be powered by two different systems, which influence both the bulb construction and its performance:

Arc lamps, are typically fabricated from clear, fused quartz tubing of a specific length. Arc lamps are closed at each end by an electrode, which is fixed with a glass to metal vacuum seal, and mounted within an elliptical or parabolic reflector enclosure. The lamp is powered by a current passing between the high voltage electrodes, creating a sustained lightning bolt (Figure 144). This is mature technology and the lamps and operating costs are relatively low. The lamps are resistant and have long life times (up to several 1,000 hours depending on cleaning intervals and switch on frequency) and are available in different lengths and with varying output [8].

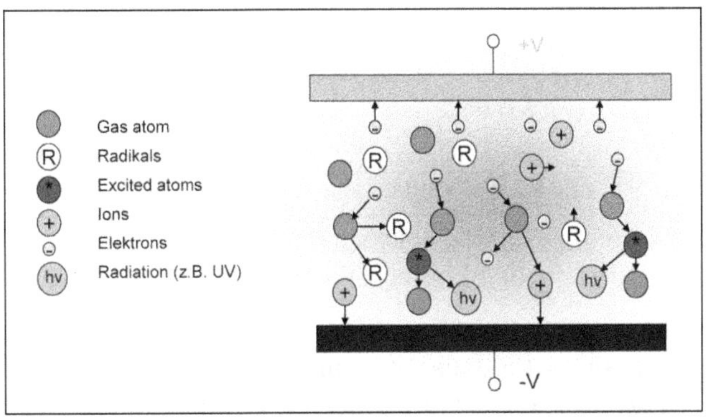

Figure 142: Composition of a plasma

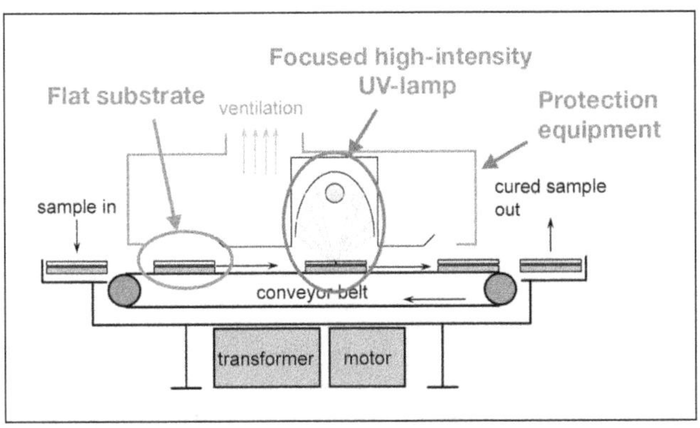

Figure 143: Schematic of a typical UV curing line

Electrode-less lamps: As its name implies, there is no direct electrical connection. Power is supplied by high powered microwaves. Power originates in a magnetron which generates a high frequency, high intensity electrical field, directing the energy down into an electrically closed cavity formed by an elliptical reflector and end reflectors, which are enclosed by a screen across the front of the lamp (Figure 145). As these lamps are without electrodes, aging due to electrode deterioration does not affect the lifetime [9].

4.1.2.2 Reflectors

Several geometrics of reflectors have been developed to meet the different requirements of diverse substrates as well as environmental conditions. Reflectors are usually shaped parabolic, elliptic or with special geometries, in order to efficiently transfer the UV light emitted by the lamp onto the substrate.

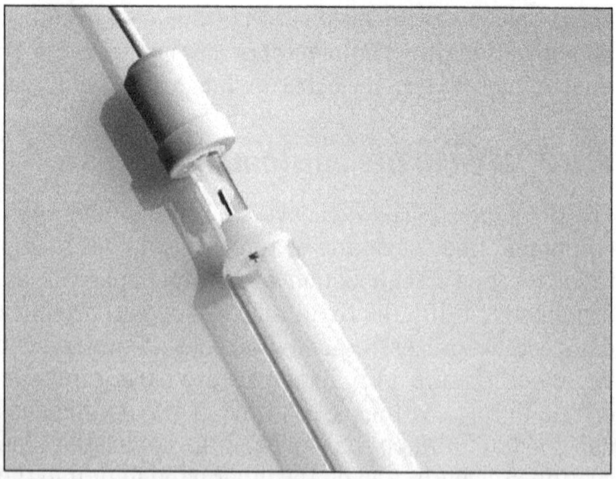

Figure 144: Arc lamp, source: IST Metz GmbH

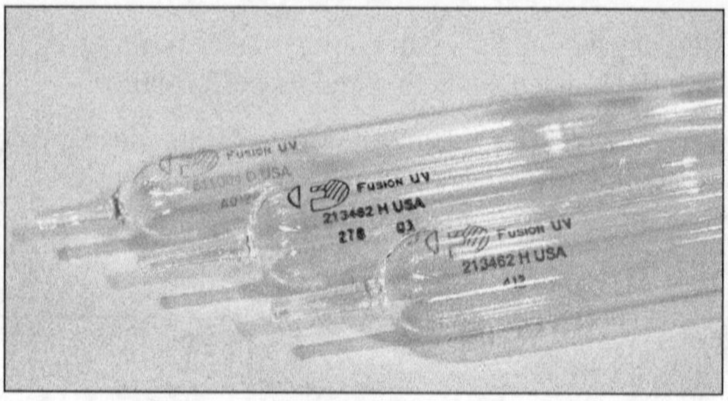

Figure 145: Electrode- less lamp, source: Fusion UV Systems

Elliptical reflectors gather and focus light energy to a narrow beam across the substrate thereby greatly increasing the intensity. The width of this beam is dependant on the bulb diameter and therefore a smaller bulb diameter results in higher intensity (Figure 146).

Parabolic reflectors gather the light energy and direct it as parallel beams to the substrate. The width of the reflector is inversely proportional to the light intensity: a broader reflector provides lower energy at any single point on the substrate (Figure 147) [10].

In addition to the geometry of the reflector, the material of which the reflector is made plays an important role. In Figure 148 the reflectivity of different reflector materials is shown.

Additionally so called cold mirror reflectors are available on the market, which are transparent to IR light. Thus the IR radiation deposed on the substrate is minimized resulting in less heating, which is important for the curing of coatings or inks applied on temperature sensitive substrates. To this end an interference coating is applied on the reflector. The reflecting

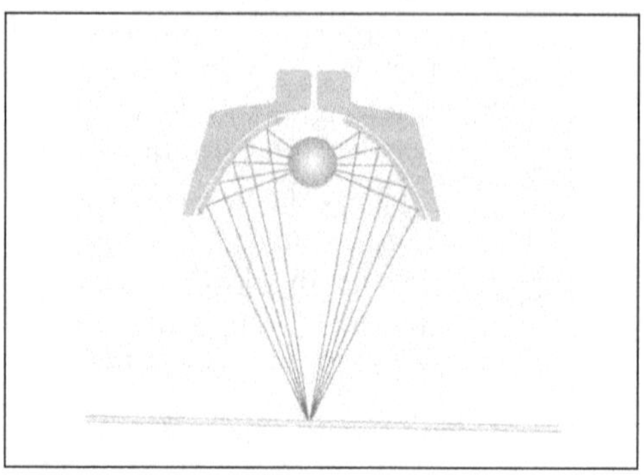

Figure 146: UV lamp reflectors with an elliptical shape

Figure 147: UV lamp reflectors with a parabolic shape

layer of these so-called dichroic reflectors is composed of approximately 50 various metal oxide layers and is highly efficient in reflecting UV light and transmitting IR radiation [11].

4.1.2.3 Inertization

As described in Chapter 2.1.1.4 oxygen inhibition is a major cause for insufficient surface cure. Curing under inert gas offers a solution to this problem. Due to advantageous costs and handling properties nitrogen and carbon dioxide are the inert gases most often used in UV curing processes (see Chapter 4.1.2.4). Nitrogen is mainly used for the curing of thin layers such as printing inks applied on 2-dimensional substrates. Under these conditions the cure speed can be dramatically increased [12 - 15].

4.1.2.4 Equipment for curing on 3-dimensional substrates

Several solutions have been developed over recent years to create solutions for curing of coatings on 3-dimensional objects. Here, the use of carbon dioxide has been used to obtain low oxygen concentrations. In contrast to nitrogen, carbon dioxide is 1.5 times heavier than air and can thus easily be used for filling even open containers with little loss to the surrounding atmosphere. The gas consumption is thus low, and the gas is provided either from cylinders or, most simply, from dry ice which produces carbon dioxide by sublimation. The carbon dioxide forms a "heavy gas lake" in which a residual oxygen content of less than 1 % can be achieved [16]. This oxygen concentration is considered to be sufficiently low for the curing of conventional free radical UV curable coatings without suffering from significant oxygen inhibition [17]. These process conditions allow the curing of coatings with lamps of lower power, opening up the possibility to use a wide range of potentially suitable lamps which emit UV light. Simple UV tanning lamps with a low intensity emission are economical and can be used. This process has been applied in the industry for the curing on lacquers on tree dimensional objects like chairs. A schematic drawing of such a pilot facility is shown in Figure 149 [18, 19].

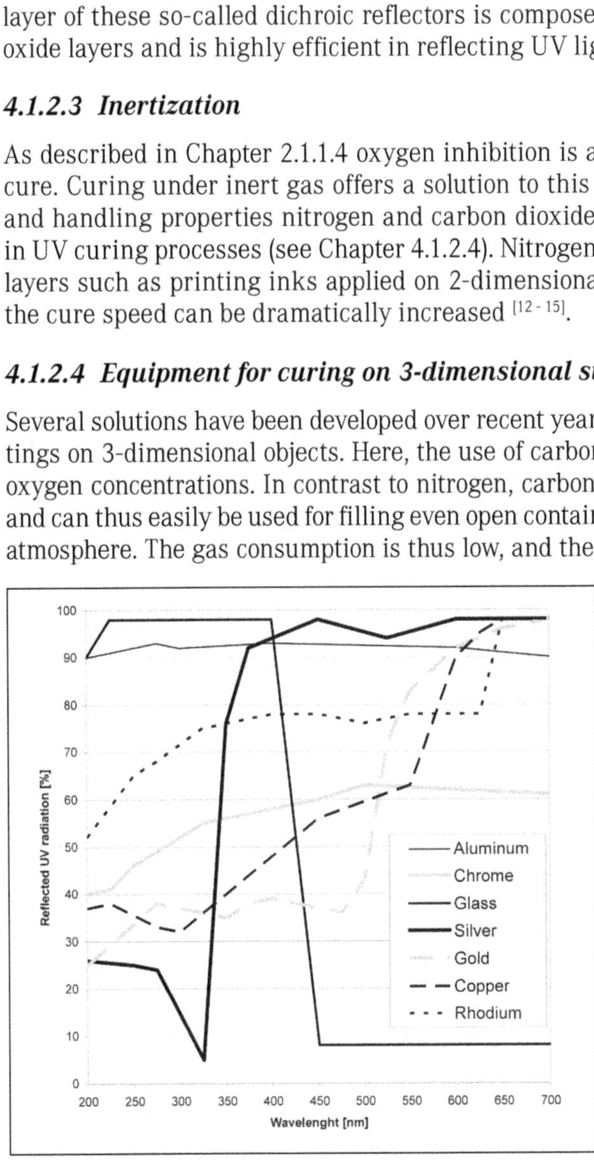

Figure 148: Different reflector materials and their UV reflection [11]

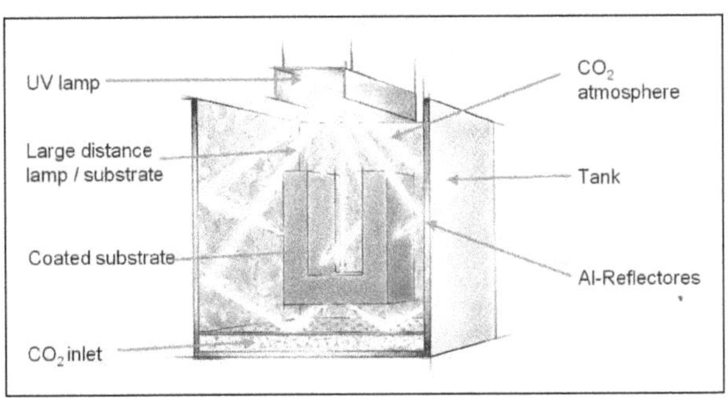

Figure 149: UV curing of coatings in carbon dioxide atmosphere [18]

In UV curing of 3-dimensional parts, a major challenge is to apply an exactly defined, homogeneous dose of UV radiation to all areas of the work piece surface. Thus, solutions have been developed which use UV lamps fixed onto robots, which then follow the shape of the object [20, 21]. In another approach to further optimize the UV curing on 3-dimensional substrates, a computer simulation has been developed which calculates the optimum relative position of UV lamps and work piece so that homogeneous UV irradiation is applied on the whole coating [22].

Figure 150: Plasma curing of a full size car body

Plasma curing technology

With the goal to light cure coatings applied on very complex shapes, like an entire car body, a further inert process has been developed. Instead of applying light from an external light source, the plasma curing technology exploits a novel concept in that the substrate is placed in a plasma chamber where UV light is generated from excited plasma. In essence, the curing is performed by placing the object to be cured into the lamp. The use of the plasma addresses on the one hand the inertization and oxygen inhibition since the curing is performed under reduced pressure in an inert gas atmosphere. On the other hand, since light is produced via excitation of the inert gas which is filling the entire vacuum chamber, UV light is produced over the entire space of the plasma chamber thereby eliminating shadow areas and making the use of reflectors unnecessesary (Figure 150).

4.1.2.5 Security

The UV curing equipment protects workers from exposure to UV light by a suitable housing. This protection is very important, because direct exposure to UV radiation results in irritation of the eyes and burning of the skin (see Chapter 7). The UV/EB protocol describes necessary measures for a safe application of the UV/EB technology. This document focuses on the significant hazards related to this technology, especially exposure to radiation. The measures described in the document are accepted by the underwriting national health and safety organizations of the European member states. With this European harmonization of measures for a safe application of UV/EB technology, the coating industry is encouraged to strive for the best use of this technology both in view of safety and environment [23].

An important element of the equipment is the elimination of ozone. Ozone is generated when oxygen is irradiated with short wavelength UVC light which is part of the emission spectra of most UV sources. This gas is toxic to humans who are able to smell it in concentrations down to 0.01 ppm, while its MAK value is defined at a concentration of 0.1 ppm (see Chapter 7). Other than elimination of ozone formed, the use of so called ozone free lamps can provide a solution to avoid exposure to ozone. In these lamps the quartz tubes contain a filter cutting off all UV light <240 nm. However, this approach has the technical disadvantage of eliminating a part of the UV spectrum that is important for many applications.

References

[1] N. Dogen, H. Klinkenberg, S. Frings, L. Reinerie, P. Wijnands, H. Andinga, 8th Nürnberg Congress Proceedings, p.155
[2] P. Mills; RadTech Europe 2005, Conference proceedings, p. 159
[3] S. B. Siegel, RadTech Europe 2003, Conference proceedings, p. 149
[4] M. N. Hirsch, H. J. Oskam, Academic Press, New York 1978
[5] W. Finkelnburg, H. Maecker, Handbuch der Physik, vol. XXII, Springer Verlag, p. 254
[6] V. Dembrovsky, Mater. Sci. Monogr., (1985), p. 23
[7] S. Santandrea;RadTech Europe 2005, Conference proceedings, p. 43
[8] B. Schwarz, RadTech Europe 1999, Conference proceedings, p. 347
[9] D. Stowe, UV Science for Non-Scientists, part I, Fusion Systems, p. 2
[10] D. Pelling, RadTech Europe 2001, Conference proceedings, p. 299
[11] J. Jung, UV-Technologie Der Praxisleitfaden für alle Druckfarben, p. 33
[12] R. Müller, RadTech Europe 2001 Conference proceedings p. 149
[13] J. Jung, P. Holl, RadTech Europe 2003, Conference proceedings, p. 498
[14] T. Henke, RadTech Europe 2001, Conference Proceeding,s p. 143
[15] K. Studer, C. Decker, E. Beck, R. Schwalm, Prog. Org. Coat., 48, 2003, p. 92
[16] E. Beck, RadTech Europe 2001, p. 643
[17] G. Bolte, Coatings, 2/2001, p. 44
[18] BASF AG Ludwigshafen (Germany)
[19] M. Biehler, E. Beck, K.Menzel, S. Titusson, A. Daiss, K. Soljamo, K. Fagerholm; RadTech Europe 2005, p. 59
[20] P. Bruger, Besser Lackieren, 15/2000, p.1
[21] U. Schmid; RadTech Europe 2001, Conference proceedings, p.53
[22] M. Schneider, W. Klein, C.Schröder; RadTech Europe 1999, Conference proceedings, p. 711
[23] B. Kütner; M. Riester; Radtech Europe 2005; Conference proceedings, p 25

5 Areas of application

The commercial applications of radiation curing coatings are very broad and growing daily. A brief list of the main applications was already mentioned in Chapter 1 of this book, but will be looked at closer on the following pages in approximate order of importance.

1. Wood coatings
2. Overprint varnishes
3. Pigmented printing inks
4. Other industrial coatings (incl. automotive coatings)
5. Electronics and telecommunication
6. Adhesives
7. Others

The main benefits and draw-backs of radiation curing formulations versus solvent-borne and also water-borne formulations in the individual application will be discussed as well. There are some general points that need not be repeated for every application (Table 28).

5.1 Wood coatings

Wood coatings account for the largest volume of radiation cured formulations on a global basis. They appear to be particularly popular in Europe, but are also used in the US and enjoy over proportional growth especially in China. The estimated volume of radiation curing wood coatings in Europe is approx. 50,000 metric tons, which equals to approx. 12 % of the total European wood coatings production.

Radiation curing wood coatings offer a large variety of technologies and formulations:
- UV curing formulations based on acrylate-functional oligomers
- UV curing formulations based on unsaturated polyesters
- water-borne UV wood coatings
- UV powder coatings

Table 28: Main benefits and draw-backs for all applications

Benefits	Draw-backs
very fast cure speed, higher line speeds possible	higher raw material cost
absence of volatile solvents (VOC), zero to low emissions of VOC	odor in case of incomplete cure
cured films display outstanding resistance properties and gloss	some raw materials are irritant
no pot-life = coating/ink does not dry before irradiation (no problems in case of machine stop; easy to clean and easy recycling)	running and maintaining the curing unit requires qualified staff

Glöckner/Jung/Struck/Studer: Radiation Curing for Coatings and Printing Inks

ISBN: 978-3-86630-907-4

5.1.1 *Conventional UV curing wood coatings based on acrylates or unsaturated polyesters*

Account for nearly 90 % of all radiation cured wood coatings. They are widely used for furniture coatings and industrial parquet coatings. Apart from the characteristics mentioned above (Table 28), they offer a variety of gloss levels and outstanding resistance properties. They are commonly applied by roller coater, curtain coater, sometimes by vacuum coating and relatively seldom by spray application.

In Europe, especially Southern Europe, the unsaturated polyesters are still very common in the wood coatings industry. Traditionally they are cured by the use of hydrogen peroxide and some cobalt octoate. Such formulations were commonly used when the new technology radiation cure appeared on the scene. By the use of photoinitiators instead of peroxide and cobalt salts they were turned into UV curable coatings. Traditionally these coatings contain styrene as monomer. Especially in Southern Europe unsaturated polyesters are still being used - primarily for fillers and sealers, but also top coats (see Chapter 3.3.2.8). The trend is to move away from the use of styrene as monomer and replacing it by less volatile and hazardous materials, such as DPGDA and TPGDA. Today step by step unsaturated polyesters are being replaced by acrylate oligomers - due to their higher curing speed and superior end use properties.

The most common acrylate-functional oligomer used in wood and furniture coatings today is polyester acrylate - combining excellent gloss, medium viscosity, good mechanical resistance and last but not least moderate cost (see Chapter 3.3.2.3). For improved flexibility and abrasion resistance in top coats polyurethane acrylates are used - often in combination with polyester acrylates (see Chapter 3.3.2.2). Epoxy acrylates are common in fillers to reduce the overall costs (see Chapter 3.3.2.1.1), polyether acrylates whenever very low viscosity is required, like for spray and curtain coating application (see Chapter 3.3.2.4).

Many UV wood coatings still contain a few percent of organic solvent that help to reduce viscosity, wet the wood pores better and improve the matting effect. Table 29 shows the main benefits and draw-backs of this technology.

5.1.2 *Water-borne UV wood coatings*

Account for approx. 10 % of the RC wood coatings market. The areas of application are mainly furniture coatings and hydro primers for industrial parquet. The water-based UV technology offers some distinct advantages over the 100 % solids technology (see Table 30 and also Chapter 3.4.1). Due to their low viscosity water-based UV wood coatings provide an excellent penetration of the wooden pores and provide excellent adhesion. They allow so called open-porous structures. They are fashionable in wood coatings to make a coated wooden surface look as natural as possible.

Table 29: Main benefits and draw-backs of RC wood coatings

Benefits	Draw-backs
immediate mounting or further processing possible	very low gloss levels difficult to obtain while maintaining rheology
coatings are very resistant towards mechanical and chemical attack	
coatings are very glossy	

Table 30: Main benefits and draw-backs of waterborne UV wood coatings

Benefits	Draw-backs
superior optical appearance, especially in case of matte and open porous finishes	additional physical drying necessary, leading to higher energy cost and slower process
formulations are free of monomers	
application equipment can be cleaned by water instead of solvent	
improved adhesion on wood (compared to 100 % solids UV)	

Moreover water-borne UV coatings are much easier to matt. Since here physical drying and water-evaporation takes place the resulting volume shrink allows the matting agent particle to work effectively (see Chapter 3.5.7 and 3.6.6). Thus, strong, reproducible, homogeneous matting effect – and if desired in combination with an open-pore structure – becomes possible. The draw-backs of this technology are the significantly longer drying times and higher energy consumption. The water has to be driven out of the film by conventional dryers, microwave dryers and/or IR lamps before UV curing.

Whereas the first generation of water-borne UV binders were plain emulsions of oligomers and monomers in water, the picture has strongly changed. The most common binders used are nowadays resin emulsions with some air-drying properties which are then crosslinked by UV to very resistant films (see Chapter 3.4.1). They outperform physically drying water-borne wood and furniture coatings with regards to chemical and stain resistance, but also mechanical resistance. Water-based UV coatings are commonly applied on wood by curtain coating, spray application or vacuum coating. Table 30 shows the main benefits and draw-backs of this technology.

5.1.3 UV powder coatings

UV powder coatings have the smallest share of coatings for wooden substrates. Currently they are not commonly applied on natural wood. Their main areas of application are coatings on MDF (medium densitiy fibre board). Overall the consumption of UV powder coatings is very low [1]. In spite of their distinct advantages like full overspray recycling and excellent resistance properties their limited storage stability and necessary temperature control during transport and storage still hinders their broad success (Table 31). Some more details about UV powder technology are described in Chapter 3.4.2.

Table 31: Main benefits and draw-backs of UV powder coatings

Benefits	Draw-backs
full overspray recycling	limited storage stability of coating powder
absence of monomers	accurate temperature control during transport and storage necessary to avoid blocking of the powder particles
excellent resistance properties	matting even more difficult than with liquid 100 % solids formulations
	flow and leveling better than in conventional powder coatings, but worse than in 100 % liquid formulations.

5.2 Overprint varnishes

Historically one of the first applications of UV curing overprint varnishes still represent one of the largest end uses of radiation curing formulations today. About a quarter of all radiation curing formulations today fall into the category overprint varnishes. As the name already indicates overprint varnishes are generally applied over printed surfaces in order to protect the printed image and to create a superior optical appearance. In most cases this is done immediately after printing at the very end of the printing machine in a so called inline process. Most common is the application on litho (offset) printed cardboard and paper substrates and in metal decoration. Actually overprint varnishes are used on nearly all kinds of inks and on all kinds of substrates. They can be applied by roller application, by gravure, flexo and screen printing. Most common applications for RC overprint varnishes are magazines, advertising brochures, cardboard boxes, labels, beverage cans, etc. Compared to solvent- and water-based overprint varnishes radiation curing systems provide higher protection and a more impressive optical effect. They can be extremely glossy or very matt. They can also feel extremely smooth and slippery. With a film weight of only 4 to 8 µm in most applications they provide superior mechanical and chemical resistance.

One special application is the partial application of overprint varnishes - often performed by silk screen printing - the so called spot coating. This technique creates very powerful effects for example by letting only certain objects in a print appear glossy or matt. The relatively high film thickness of the screen applied varnish makes it possible to really feel the coating when touching such a surface.

During the last decade radiation curing overprint varnishes have moved from an exclusive coating to very standard formulations which experience a strong price competition in the market. There are certainly special applications in which a high price level can be maintained, but the bulk of UV overprint varnishes has a significantly lower price level today compared to several years ago. The volumes keep continuously growing.

Overprint varnishes need to wet the freshly applied ink layers readily. They have to cure fast and completely, the cured film needs to be scratch resistant and ought to display any desired gloss level on demand. Further benefits and draw-backs are summarized in Table 32.

Besides being economically advantageous epoxy acrylate oligomers provide excellent reactivity and hardness, which makes them the first choice for most overprint varnish formulations. For coatings on flexible substrates they are too brittle and do not provide the necessary adhesion (see Chapter 3.3.2.1). Often polyurethane acrylates in combination with others are used to fulfil the requirements (see Chapter 3.3.2.2).

Table 32: Main benefits and draw-backs of UV curable overprint varnishes

Benefits	Draw-backs
superior resistance properties	very low gloss levels difficult to obtain while maintaining rheology
superior finish and gloss	
strong effects achievable	
varnishes maintain their properties during long print runs since no solvent or amine evaporates	

5.3 Pigmented printing inks

In the graphic arts industry there is a relatively clear and easy distinction between inks and varnishes: inks are colored (and generally pigmented) whereas varnishes are non-pigmented. The pigments used are generally organic pigments and carbon black pigments. Compared to coatings the pigment loadings in printing inks are rather high since inks must transfer a distinct color impression with only a few micrometers film thickness. The use of such "highly absorptive" pigments at high levels adversely affects the curing properties of a UV-curing ink.

Consequently in the early days UV curing of pigmented inks was very rare. In graphic arts radiation curing started off with varnishes. After the development of tailor-made photoinitiators (see Chapter 3.2.5) and more powerful UV lamps (see Chapter 4) pigmented inks became popular. They have been growing over-proportionally during the last years. However they still only contribute to less than 5 % of the global printing ink production – but with respectable growth rates.

A special aspect to be discussed with regards to RC inks is their potential use in food packaging. In food packaging only a very small amount of extractable (assumingly identical with migratable) substances are accepted in the cured film. But even the presence of photoinitiators does not prevent the use of UV inks in indirect food contact. By the use of polymeric and/or crosslinkable photoinitiators formulators manage to stay below the required level of extractables [2]. Here is a clear advantage for the use of EB curing, since in these inks no photoinitiators are used.

Pigmented RC inks are applied by nearly any standard printing technique – except for gravure printing due to viscosity constraints. Radiation curing has gained a large market share in the area of screen inks, a much smaller share with the others. Overproportional growth for RC inks can be observed with any process.

5.3.1 Screen printing

Screen printing is also called silk screen printing or serigraphie. It employs a stencil principle. A rubber squeegee is used to push ink through the stencil (screen) on to the substrate [3]. The design-bearing fabric is mounted on a metal frame, the material used today most commonly is nylon, historically it used to be silk. The non-printing areas of the screen fabric are

Table 33: Main benefits and draw-backs of RC inks

Benefits	Draw-backs
inks do not dry on the roller or in the screen (no recovering and cleaning needed after machine stops)	adhesion may be more critical than with solvent-based screen inks
inks maintain their properties during long print runs since no solvent or amine evaporates	higher raw material cost than with solvent-borne inks
excellent chemical and mechanical resistance properties	
suitable for a huge variety of substrates (plastic, film, glass, metal, etc.)	
superior print quality, fine lines and gloss	
immediate processing possible due to fast cure	

blocked by a photopolymer, the open areas form the image to be printed. Inks used in the screen printing process display a rather high viscosity in the range of 1.5 to 6 Pa s under shear, often going along with some thixotropic flow behavior. Figure 151 shows the principle and Figure 152 of an automated lab screen printing unit for flatbed printing.

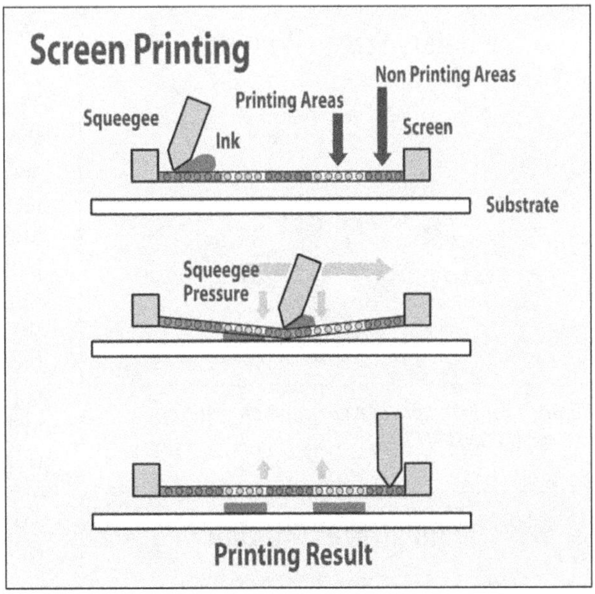

Figure 151: Principle of screen printing, source: EVONIK TEGO

Screen printing inks were the first pigmented inks cured by radiation on an industrial scale about 30 years ago. Today (2008) about one third of all screen inks are radiation curing. Besides the pigmented screen inks many varnishes (non pigmented) are applied by screen printing with full coverage or as a so called spot lacquer.

In the screen printing process significantly higher film weights are applied compared to other printing processes. It is generally used to print all kinds of substrates and to create prints that are very resistant and/or provide a very powerful visual impression. Substrates of any format can be screen printed. The most common areas of application are:

- labeling of CDs and DVDs
- posters and large format displays
- plastic containers and labels
- point of sale displays

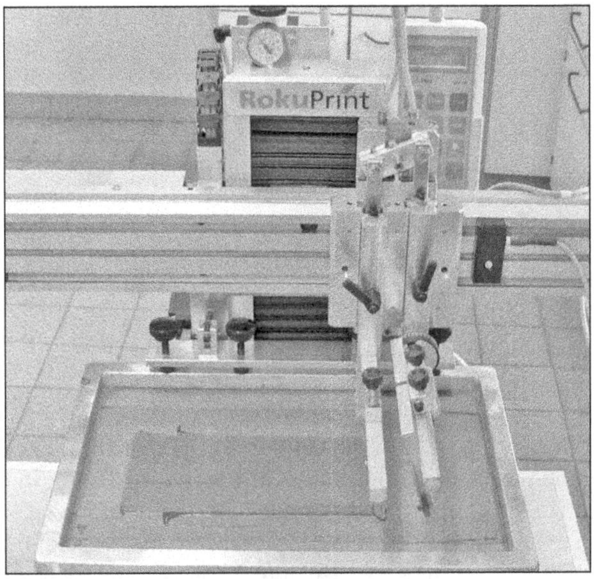

Figure 152: Automated silk screen printing lab unit, source: EVONIK TEGO

Radiation curing screen inks are used in flat-screen printing on any kind of flexible to rigid substrate and as rotary screen printing on web fed flexible substrates. Rotary screen printing uses hollow screen-mesh cylinders and squeegee blades located inside the cylinders that force the ink onto the substrate as it travels through a printing press. It is the ideal method for printing wallpapers and labels with possible film weights >>10 µm.

Silk screen printing is highly versatile, and is used in applications outside traditional printing, such as mask application in the electronics industry and surface decoration of plastic consumer goods and automotive components. The oligomer selection for screen inks depends primarily on the substrate to be printed and on the necessary resistance properties of the cured inks and varnishes. It nearly covers the full range of oligomers – from epoxy to polyurethane acrylate.

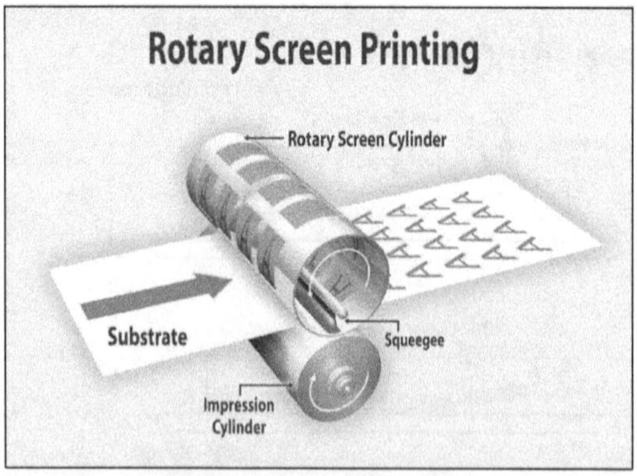

Figure 153: Principle of rotary screen printing, source: EVONIK TEGO

Figure 154: Principle of a lithographic printing unit, source: EVONIK TEGO

5.3.2 Litho-offset

Lithography is a planographic process. The image and non-image areas are in the same plane on the printing image carrier [4]. Lithography makes use of the different wetting properties on hydrophilic and hydrophobic surfaces. The printing plate has distinct hydrophobic and hydrophilic areas. During the printing process the hydrophilic areas are wetted by fountain solution (water, surfactant, additives, commonly also alcohol), the hydrophobic areas are then wetted by the oily ink (Figure 154). Since these printing plates are not resistant to abrasion the image is transferred to a rubber blanket and from there to the substrate (offset).

Globally offset is the largest printing process. When offset-printing is done on paper or cardboard sheets this relatively slow printing process is called sheet-fed offset. The conventional inks are air-drying by penetration and evaporation of mineral oil and by slow chemical cross-linking induced by the oxygen out of the air. After printing the sheets are often powdered to avoid blocking in the stack of printed sheets. It takes some time until further processing is possible. UV curing inks achieve excellent print qualities, make the powder obsolete (higher gloss) and allow immediate further processing.

When printing is performed on a paper or film web (material coming from a roll) this is called web offset. These inks are normally drying in thermal drying ovens. The use of UV curing inks eliminates the need for drying tunnels, saves space and energy cost. Whereas conventional offset inks are nearly exclusively applied on paper and cardboard, radiation curing offset inks are also applied on various kinds of film.

Overall the highest volumes of pigmented RC inks are UV litho inks. Main applications for sheet-fed:

- packaging
- commercial sector for high value brochures, direct mail
- currency paper, bank notes
- credit cards, telephone cards, identity papers

Main applications for web offset:

- forms and direct mail
- cheques and lottery tickets
- labels
- packaging

5.3.3 *Flexographic printing*

Flexographic printing is a relief printing process and derived from letterpress. But instead of solid typefaces the printing plate is made of flexible photopolymeric material, whose elevations transfer the ink directly to the substrate. The inks used in the flexographic printing process are liquid and their viscosity is traditionally very low. Solvent-based and water-based flexo inks have an efflux time in a DIN 4 cup between 16 and 28 seconds, which equals to approximately 50 to 100 mPa s. With radiation curing raw materials however this viscosity level cannot be achieved. Figure 156 shows the principle schematically and Figure 157 a flexo printing plate.

Figure 155: Closeup of a narrow-web lihto printing unit, source: Nilpeter

The volume of RC flexo inks globally is still relatively small, but growing with double digit rates. Latest developments of low viscosity oligomers and pigment wetting components have further boosted the performance of these inks. Besides digital inks flexographic inks are the fastest growing segment of radiation curing inks. There are UV curing and meanwhile also EB curing ink systems available [5].

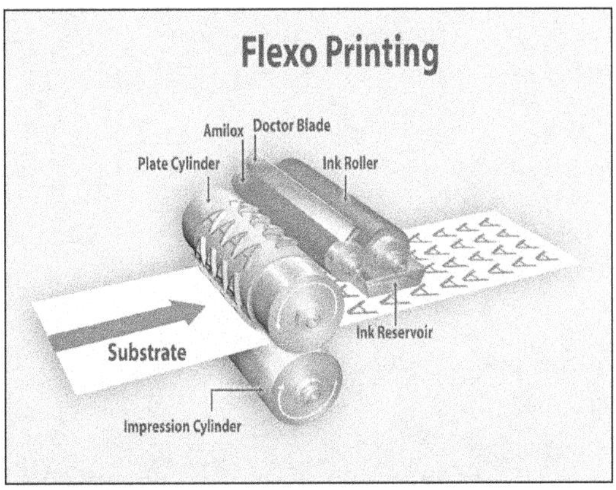

Figure 156: Principle of flexo printing, source: EVONIK TEGO

Radiation curing flexo inks can be printed on various substrates - from paper and cardboard to foil and film. The cured prints are extremely sharp and provide excellent mechanical and chemical resistance. They are applied both on wide web and on narrow web. When printed on wide web all kinds of substrates are printed - generally for high quality packaging applications. Narrow web application is very typical for the largest use area of these inks - label printing. Usually this is not solely done by flexo printing, but in a so called combination printing process. The industrial label printing machines are equipped in a way that each printing unit can be used for different printing techniques. It is very common to combine screen printing with flexographic printing and possibly letterpress printing on one single machine. Figure158 shows a modern label printing machine.

5.3.4 *Letterpress*

Letterpress is the classical relief printing process. The image area is raised above the non-printing areas, is inked by rollers and pressed into contact with the substrate [4]. Letterpress inks display very high viscosities and are commonly called paste inks.

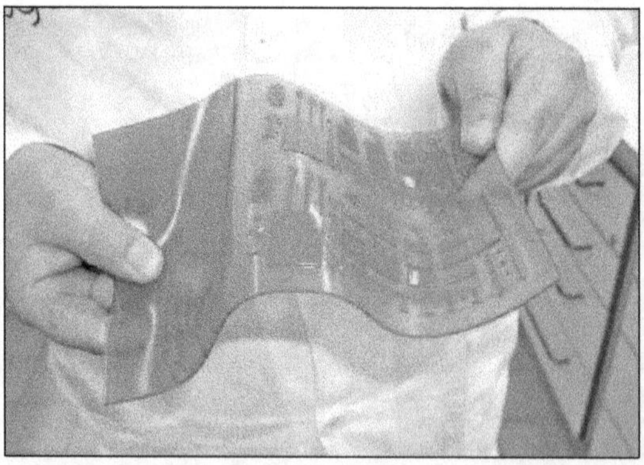

Figure 157: Flexo printing plate, source: EVONIK TEGO

Figure 158: Picture of a high end narrow press, eg. for label printing, source: Nilpeter

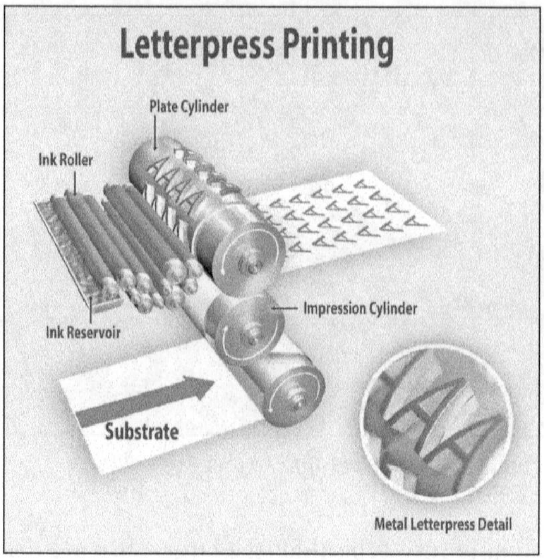

Figure 159: Principle of letterpress printing, source: EVONIK TEGO

UV letterpress is predominantly used in the packaging and label industry as direct and indirect printing. Indirect printing means the typeface prints the mirrored picture onto a rubber surface, which finally transfers the image to the substrate. Indirect printing is used to print objects that could not be printed directly because of their shape (e.g., yoghurt cups). It is also quite popular in metal decoration.

The total volumes of UV letterpress inks are rather small. Letterpress as a process in general is extremely small making up for only 1 or 2 % of the global printing ink production. And only a part of this is radiation curing.

5.3.5 Industrial inkjet printing

Inkjet printing is a digital printing process. This means that in order to print it is not necessary to prepare a printing plate or screen. The data for the image to be printed comes from a digital source (computer) and are transferred into an image directly. In the case of inkjet printing this is done by inkjet heads that jet ink droplets to the substrate in a very controlled way at enormous speed. If needed the data can be changed for every print, which makes this technology ideal for coding, marking and personalization. UV ink technology is used in both of the main inkjet technology types, both continuous inkjet (CIJ) printing and in drop on demand or piezo (DOD) inkjet. But it is DOD printing which has exploited to full potential of the technology and is enjoying enormous growth. CIJ printing is mainly used for coding and marking of articles from food packaging, where date codes are required, to automotive where very resistant batch and identification numbers are required to survive the life of the vehicle. The formulations are often cationically curing.

It is standard to use piezo print heads for dot on demand (DOD) printing with radically curing radiation curing inks. In a piezoelectric inkjet printer, small walls in the nozzle filling chambers are made of piezoelectric material that changes shape (deflects) when a voltage is applied. The deflection of the walls forces a drop of ink out of the nozzle (Figure 160).

Table 34: Main benefits and draw-backs of UV curing printing inks

Benefits	Draw-backs
outstanding print quality and gloss	more costly than solvent-based or water-based inks
no drying on print cylinders in case of machine stop, less cleaning	
can be applied to temperature-sensitive substrates, such as film	
inks maintain their properties during long print runs since no solvent or amine evaporates	
excellent resistance properties	

RC formulations for ink jet printing are the latest addition to the range of radiation curing printing inks. Although a prime area of R&D on the equipment and formulation side, the volumes sold on a global basis are actually still small at present. The current volume of industrially applied UV jet inks is estimated to be between 1,000 and 2,000 tons per year. However growth rates of 20 to 100 % per year depending on the application will change this picture continuously.

UV inkjet is currently being tested and applied in various applications. This technology serves the need for printing on demand, personalization and short print runs. In many applications it competes with pad and screen printing application - especially when only a small number of prints needs to be made. Then the savings on screen preparation economically make up for the lower print speed. UV jet inks are even used to print on leather and artificial leather to create patterns with excellent optical and resistance properties. In other applications there are no alternatives to ink jet printing. The possibilities of personalization, date/time stamping and numbering offered by this digital printing process offer numerous new applications. Currently the main applications for UV jet inks are large format printing (e.g., for posters and displays) and coding and marking. Figure 161 shows an example of flat bed ink jet printer utilizing UV curing.

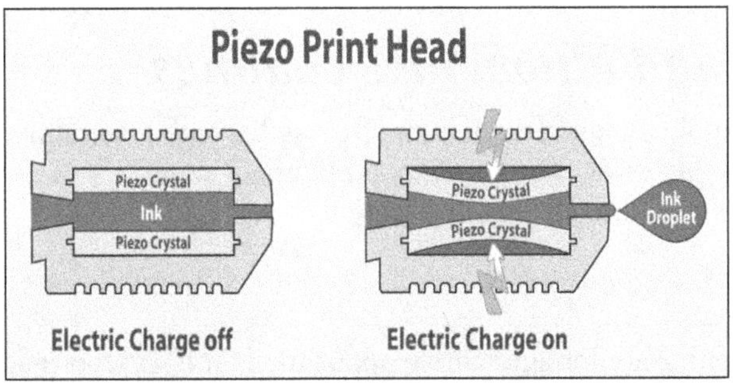

Figure 160: Principle of a piezoelectric print head, source: EVONIK TEGO

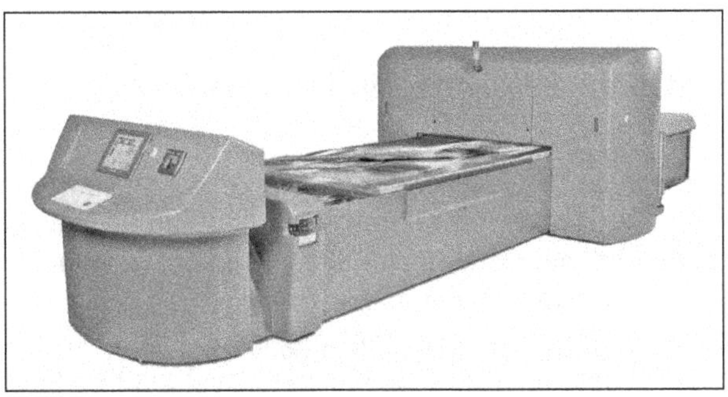

Figure 161: Flat bed UV ink jet printer, source: Fujifilm Sericol

Besides the fast cure and the excellent resistance properties of the cured inks the big advantage of using UV inks is that they do not dry by evaporation and do not tend to clog the nozzles of the jet head. Most formulations are radically curing and free of volatile organic solvents.

Table 35: Main benefits and draw-backs Industrial inkjet printing

Benefits	Draw-backs
printing on demand (no plate or screen preparation)	relatively low print speed
small runs more economical than with e.g. screen printing	less economical than other printing processes in case of multiple copies
wide formats possible	
direct printing on any substrate possible	
lower print head maintenance (compared to solvent and water-based jet inks)	
no nozzle clogging by drying ink, less down-times (compared to solvent and water-based jet inks)	
superior resistance properties (compared to solvent and water-based jet inks)	

The inherent viscosity of acrylic UV oligomers is an obstacle for ink jet printing since very low viscosities (8 to 20 mPa·s at the jetting temperature) are required. Therefore, the print heads are heated up to 60 °C in order to reduce ink viscosity for jetting. Standard oligomers find limited use in jet inks. Unlike in other ink formulations often high amounts of mono- and difunctional monomers are used because of their low viscosity. Combinations with tri- or tetrafunctional acrylates ensure sufficient cure speed and resistance properties [6]. Table 35 lists the main benefits and draw-backs of this technology.

5.4 Other industrial and automotive coatings

Whereas in wood and printing ink applications radiation curing has become a standard technology this is only true to a much lesser degree for many other applications. Table 36 indicates how widely RC technology is used in different applications.

5.4.1 Automotive coatings

When discussing radiation curing technology for automotive applications it is appropriate to distinguish between

- automotive parts
- OEM body coating
- car repair coating

Overall the volumes of UV coating used in the automotive industry are still very small, accurate numbers cannot be quoted.

A lot of research has been and is still being done on automotive applications, many new raw materials and curing equipment developed for this application [7, 6]. Without any doubt RC coatings have reached a performance level that makes them suitable for the use in automotive applications. Compared to water-based and even solvent-based coatings up-to-date UV formulations provide superior mechanical and chemical resistance properties, equal optical appearance and nearly equal adhesion. Curing on three-dimensional objects is possible, but still relatively complex. Some coatings for certain automotive parts have already been a standard in the industry for several years [8].

Table 36: Frequency of use of RC technology in different industrial applications

	First tests	Line trials	First industrial use	Wide industrial use
Certain automotive parts				
OEM coating				
Car repair				
Coil coatings			2007	
Adhesives				
Electronics and telecommunication				
Silicone release coatings				

Automotive headlenses

The specific automotive application where UV has become the standard technology is in the automotive forward lighting industry, where UV coatings have been used for more than 15 years and now command 80 % of the market. Headlamps are composed of two primary components that need to be coated – the polycarbonate lens and the reflector housing. The lens requires a very hard, scratch-resistant coating to protect the polycarbonate from the elements and physical wearing. The reflector housing has an UV basecoat (primer) that seals the substrate and provides an ultra-smooth surface for metallization. The reflector basecoat market is now essentially 100 % UV cured. Most of these coatings do still contain a certain percentage of organic solvents today.

Exterior plastic parts

One of the lesser known applications is the use of an UV curable clear coat over molded-in-color body side moldings. The coating has to be very tough and flexible to maintain adhesion without cracking from objects striking the molding. The drivers for the use of UV coatings in this application are the speed of cure and superior performance properties.

SMC body panels

Sheet molding compound (SMC) is a composite material that has been used as an alternative to steel for more than 30 years. SMC consists of a glass-fiber-filled polyester resin that has been cast into sheets. These sheets are then placed in a compression mold and formed into body panels. However, these are porous substrates. When coated a paint defect known as a “porosity pop” can occur. To prevent this using of a dual-cure (UV/thermal) sealer has become standard in some coating lines.

Alloy wheels

Alloy wheels are commonly being coated by a solvent-based 2-pack coating, requiring intensive drying and cure in thermal ovens. There are first applications in the industry employing UV powder (but also liquid 100 % UV). Even though some heating is still required the resulting energy and time savings in the process are very high.

OEM clearcoat

UV clear coats are still not a standard in the automotive industry. Radiation cure of automotive OEM coatings on the car body has been a very complex challenge that could only be mastered by the joint effort of coating manufacturers, curing equipment manufacturers, car producers and raw material suppliers. In principle UV cured automotive clear coats are an option today.

Table 37: Main benefits and draw-backs of UV curable automotive coatings

Benefits	Draw-backs
high savings on energy cost and CO_2 emissions	higher raw material costs
reduced space requirements for coating line	no or limited cure in shadow areas
superior scratch resistance	dual-cure technology supplies some problem solutions, but limits benefits of UV technology
excellent chemical resistance	

It has been impressively demonstrated in several publications that UV clears exhibit superior scratch resistance and chemical resistance compared to the currently used solvent-based top coats [9]. Thanks to major advances in stabilizer technology this is possible without sacrificing UV and weathering resistance.

However a draw-back for UV technology on 3-dimensional objects like car bodies is the lack of cure in shadow areas. This is why the dual-cure currently appears to be the technology of choice (see Chapter 3.4.3). It combines a basic level of standard isocyanate crosslinking with the high crosslink density of radiation curing formulations. If there is not enough UV radiation in shadow areas at least the coating crosslinks to an acceptable level. In order to overcome the poor crosslink density in shadow areas several new cure facilities have been developed recently (see Chapter 4.1.1.5).

Both DuPont and BASF have installed pilot lines at their facilities. Not only do the coating companies have to show good coating performance, they also have to demonstrate a paint-line solution. One of the other benefits of UV/thermal curing cited by DuPont is that the length of the finishing line can be reduced by 50 % simply by reducing the length of the thermal oven. It is not known if and when UV clears will become a standard in the automotive OEM industry. Their implementation into a car coating line means a major financial investment. But they are definitely an option to consider when designing a new automotive coating line.

Car repair

Stronger use of radiation curing coatings could make car repair coatings much easier and quicker. For repairing minor damages, the so called spot repair, it has become more and more common to utilize UV cured primers. The curing equipment used today is based on flash light technology. The UV flash cure unit is compact, designed for spot repair applications and emits UV light over a wide wavelength range. The coating is cured in less than half a minute. It is very likely that the other coatings layers (filler, basecoat and top coat) will follow in the future. Recent publications discuss UV-A curable clear coatings for spot repairs up to total resprays [10]. Table 37 shows the main benefits and draw-backs of UV curable automotive coatings.

5.4.2 Coil coatings

In principle the coil coating application ought to be ideal for radiation curing. A flat metal coil is coated by roller application at rather low film thickness, cured and processed at high speed. The resulting coatings are very durable and resistant.

Today all common coil coating applications work with solvent-based formulations. Curing respectively drying is achieved in drying tunnels in less than one minute at temperatures of above 200 °C. Substrates used are steel, galvanized steel and aluminum. Coil coating lines are up to 2.30 m wide and run line speeds up to 200 m/min.

Main areas of application are

- construction (wall panels, roofing, etc.)
- transportation (panels for use in automotive body panels, etc.)
- household appliances (washing machines, refrigerators, computers, etc.)

A detailed breackdown is shown in Figure 162.

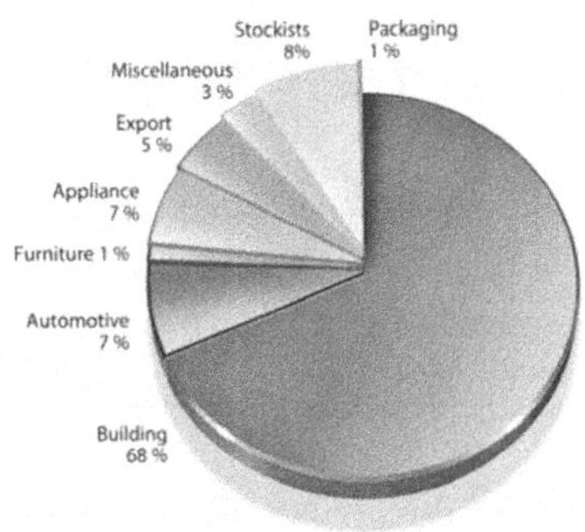

Figure 162: Main coil coatings markets in 2005 (steel + aluminum), source: ECCA

The application of UV or EB for coil coatings in North America and Europe is fast gaining recognition and represents a significant potential market for the technology [11]. Many sources report about trials or smaller uses of water-based, powder or radiation curing formulations. But until now there have been obstacles preventing the break-through of one of these: Water-based coatings increase drying times and energy costs dramatically. The process would not be economic any more. Powder coatings do not display the desired homogenous flow and are difficult to apply at low film thicknesses.

In the last few years a lot of work has been invested to overcome the obstacles of radiation curing formulations for the use in coil coating lines, which are:

- poor adhesion to metal substrates
- insufficient flexibility
- through-cure problems due to high pigmentation (especially with TiO_2)

Latest developments on the raw material side have overcome these problems [12]. Today it is possible to apply and cure a UV curing coil coating with excellent properties. Several trials have been successful. The first large scale application is just being built in Belgium by Arcelor. Table 38 shows benefits and draw-backs of RC coil coatings.

5.5 Adhesives

The radiation curing technology is not only used to cure surface coatings and inks, but is also highly suitable for adhesives. With the move towards solvent-free technologies UV curing adhesives have emerged in the industry. Their use is mainly limited to applications in which transparent materials are used or in which the adhesive is applied as a surface coating.

Main areas of application for radiation curing adhesives are

- laminating adhesives
- pressure sensitive adhesives on label backsides and adhesive tapes
- bonding of optical discs (CDs and DVDs)

Table 38: Main benefits and draw-backs of RC coil coatings

Benefits	Draw-backs
enormous savings on CO_2 emissions	higher raw material costs
reduced space requirements for coating line	lamp settings have to be adjusted according to specific color or formulation
savings on energy costs	still a very young technology

Table 39: Main benefits and draw-backs of RC adhesives

Benefits	Draw-backs
tack on demand	limited to surface coatings or transparent materials
excellent optical clarity of the cured film (optical disks, laminates)	special formulation know-how necessary to finely control tack and bond strength
energy and time savings (no heating, faster application lines)	

The overall market today is still rather small, but growing with double digit figures. Global studies indicate that 1 to 2 % of all radiation curing formulations fall into the category adhesives. According to US Radtech figures adhesives represented nearly 4 % of the US radiation curing market in 2005 [13].

Adhesives are formulated differently compared to coatings or inks. The main differences in formulation and application are:

- lower degree of functionality on oligomers and monomers
- lower degree of crosslink
- lower cure speed, but accurate control of polymer architecture
- hardly any pigmented formulations
- film thickness often higher than with most coatings and inks

Table 39 shows main benefits and draw-backs of RC adhesives.

5.6 Electronics and telecommunication

Radiation curing systems for the use in electronic and telecommunication goods account for more than one tenth of the global production of RC formulations. The main uses of UV curable coatings in electronics and telecoms are:

- printed circuit board (PCB) and printed wiring board (PWB) assembly
- coatings on optical fibres

PCB and PCW

In electronics, printed circuit boards (PCB) are used to mechanically support and electrically connect electronic components using conductive pathways, or traces, etched from copper sheets laminated onto a non-conductive substrate (Figure 163).

Most PCBs are composed of between one and twenty-four conductive layers separated and supported by layers of insulating material (substrates) laminated (glued with heat, pressure and sometimes vacuum) together. The vast majority of printed circuit boards are made by adhering a layer of copper over the entire substrate, sometimes on both sides, (creating a "blank PCB") then removing unwanted copper after applying a temporary mask (e.g. by etching), leaving only the desired copper traces. The most common method to control the removal of copper is silk screen printing. The printing process applies etch-resistant inks to protect the copper foil. Subsequent etching removes the unwanted copper. Alternatively, the ink may be con-

Figure 163: Photograph of a printed circuit board, source: EVONIK TEGO

ductive, printed on a blank (non-conductive) board. Areas that should not be soldered to may be covered with a polymer solder resist (solder mask) coating. The solder resist prevents solder from bridging between conductors and thereby creating short circuits. Solder resist also provides some protection from the environment. Line art and text may be printed onto the outer surfaces of a PCB by screen printing. When space permits, the screen print text can indicate component designators, switch setting requirements, test points, and other features helpful in assembling, testing, and servicing the circuit board. The upstream PCB coatings market, consisting mainly of solder masks, etch resists and plate resists, is now almost entirely UV curing. The thermally cured screen-printed coatings that compete in this market are gradually being phased out.

Optical fiber coatings

An optical fiber is a single, hair-fine filament drawn from molten silica glass. These fibers are replacing metal wire as the transmission medium in high-speed, high-capacity communications systems that convert information into light, which is then transmitted via fiber optic cable.

In a fiber optic cable, many individual optical fibers are bound together around a central steel cable or high-strength plastic carrier for support. This core is then covered with protective layers of materials such as aluminum, "Kevlar", and polyethylene (the cladding). Because the core and the cladding are constructed of slightly differing materials, light travels through them at different speeds. As a light wave traveling in the fiber core reaches the boundary between the core and cladding, these compositional differences between the two cause the light wave to bend back into the core. Thus, as a pulse of light travels through an optical fiber, it is constantly bouncing away from the cladding.

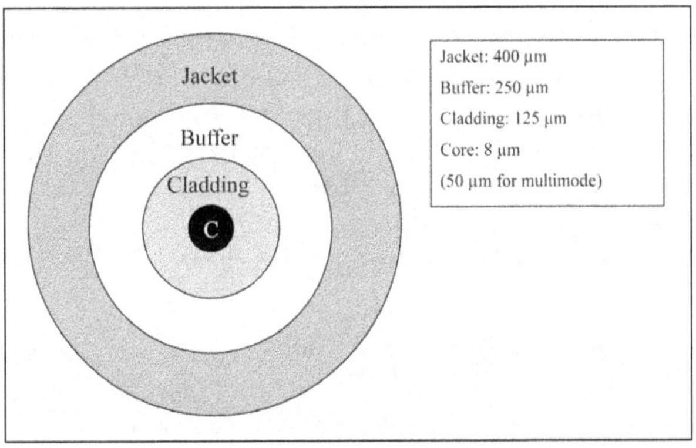

Figure 164: Principle of a fiber optic cable

On-line coating during fiber drawing is essential for protection of the fiber surface. To achieve sufficiently high fiber-drawing speeds the coating must be applied and solidified in a very short time. The viscosity of the coating is preferably in the order of 1 Pa·s at the application temperature. For these reasons UV curable acrylates are the preferred class of coating materials [14].

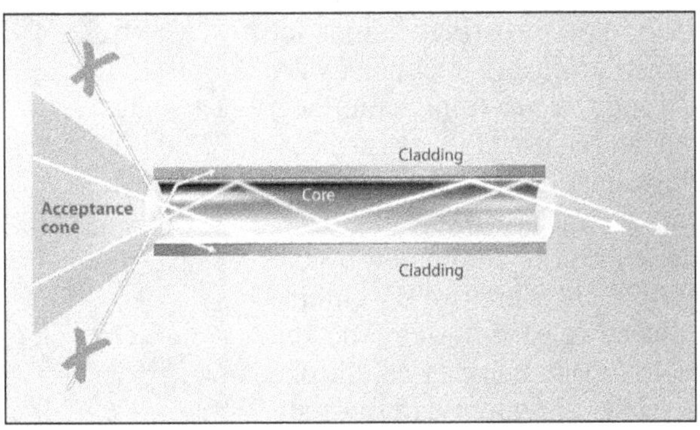

Figure 165: Guidance of light through an optical fiber, source: EVONIK TEGO

A very important function of the coating is to protect the fiber from micro-bending losses. Such small period distortions may be caused by lateral forces, for instance when the fiber is wound on a drum under tension or when it is mounted in a cable. Microbending results in transmission losses when the optical modes leave the core. In order to decrease the micro-bending losses the fiber is surrounded by a dual coating system consisting of a soft buffer and a hard top layer. The soft coating provides the soft compliant enclosure which decouples the fiber mechanically from its environment. The hard top coating acts as a stiffener which withstands external forces. These functions must be performed over a wide temperature range, preferably from -60 to +80 °C. Today these challenging requirement are fully met by radiation curing formulations. They appear to fully cover the market of primary and secondary coatings for optical fibers.

5.7 Others

5.7.1 Radiation curable silicone release coatings

Another field of application where radiation curing has gained a significant market share are silicone release coatings on papers and films. The main end uses are

- carrier papers (release liners) for pressure sensitive labels (PSA labels)
- release coatings for various adhesive tape constructions, and
- release coatings in the field of feminine and baby hygiene applications (e.g., diapers), see Figure 166

The current global market for silicone release coatings is estimated to be approx. 26,000 metric tons of thermally curing solvent-based/solvent-less products, and radiation curing release systems. In 2006 the worldwide demand for siliconized substrates (release liners) was 30.5 billion m² [15]. Silicone release coatings for paper and filmic substrates can be crosslinked by heat cure and by "cold" cure [16].

The traditional thermal cure can be achieved by a condensation or addition polymerization process. In the condensation process an OH difunctional silicone is cross-linked usually by an alkoxy-, alcyloxy -or hydrogen-functional silicone compound – often in the presence of a tin catalyst. In the curing process, however, the main reaction is the crosslinking of a vinyl-functional silicone with a hydrogen-functional silicone catalyzed using platinum group derivatives. The heat applied in both processes (above 100 °C) limits their use to the less heat sensitive substrates, such as papers and polyester films.

Figure 166: Typical applications for RC silicone release coatings

The term "cold cure" with silicone release coatings stands for radiation curing technology by UV light or electron beam. Even though the use of RC release coatings started off in 1980s with EB cure technology in the US, today the market is dominated by UV curing. There are two UV curable silicone release systems on the market. One is based on silicone acrylates and cures via a free radical mechanism, while the other system uses epoxy silicones and

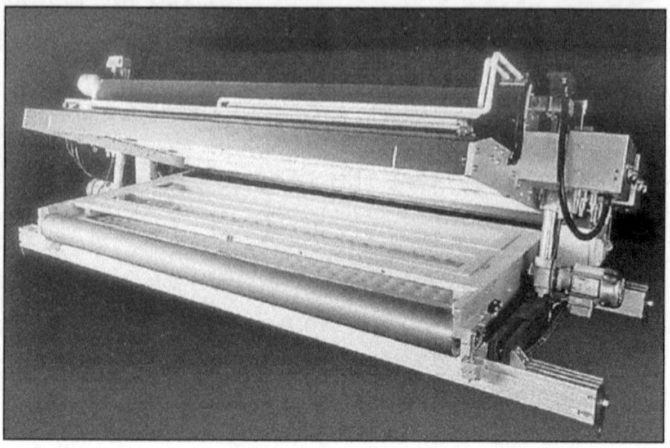

Figure 168: Inerted UV unit for curing silicone acrylates via free radical mechanism, source EVONIK Goldschmidt

Figure 168: 5-roll coating head for UV curable silicones, source: EVONIK Goldschmidt

cures in the presence of cationic initiators. Both require exposure to ultra-violet light to initiate the curing reaction; silicone acrylates can also be cured by EB. Both radiation curing systems are supplied and used without solvent.

Cationic curing differs from free radical curing because it can take place without inertization. However, this system undergoes post-curing because of the "slow" cationic curing reaction. Intended inline conversion must be checked very carefully. Because of the chemical sensitivity for alkaline components being present as additives in several papers, films or top coatings the use of cationic release coatings is limited to well-chosen substrates. In contrast, the free radical curing silicone acrylates are very robust. The silicone coating is ready for conversion, in-line adhesive coating or quality testing right after the inerted UV unit. They can be used both on today's standard base papers and films. Inerting is necessary to avoid the inhibition of the cross-link reaction by oxygen.

The viscosities of the RC release coatings range typically from 100 to 2,000 mPa·s. The release properties of the silicone acrylates and epoxies typically used are determined by their chain length, number of functional groups per molecule, and the spacer element between siloxane backbone and functional group (see Chapter 3.3.2.5). By mixing different silicone acrylates or epoxies it is possible to fine tune the release properties; release values between premium easy till super tight are possible, depending on the application. The liquid release coatings are applied via a 5-smooth-roll coater head, or a 3-roll offset-gravure coating system (Figure 168). For a sufficient coating quality 0.8 to 1.5 g of silicone per m^2 are needed on a typical release liner substrate.

RC silicones can be used on a large variety of substrates. Thermal sensitive substrates, e.g., polyolefinic films, thermo papers, are the preferred materials for silicones which cure at ambient temperature. LDPE, HDPE, and thin boPP films (e.g., 30 μm) are common substrates for release liners for label and hygiene applications. With paper liners the loss of water during the polymerization process of thermally curing silicones is avoided; rehumidification is not necessary, and the siliconized paper has excellent lay-flat properties. Table 40 shows main benefits and draw-backs of this technology.

5.7.2 Radiation curing scratch-resistant coatings

A highly interesting approach in the industry is to combine the principles of nanotechnology and radiation curing in order to achieve highly scratch resistant yet transparent

Table 40: Main benefits and draw-backs

Benefits	Draw-backs
coating of heat sensitive substrates possible	inertization required for silicone acrylates
high curing speed	more expensive than thermal silicones
space and energy savings due to the absence of thermal drying tunnels	post-curing with silicone epoxies

Figure 169: UV clear after scratch/abrasion test, left: without silica, right: with 16 % silica, source: EVONIK Degussa

coatings. (Possible) Applications for this approach are

- coatings for TFT displays
- wear-resistant panels for furniture
- automotive clears
- numerous others

For this purpose specially treated silicas with special morphology are used that can be incorporated into the radiation cured films up to 20 %. They act as functional fillers and dramatically improve the scratch and abrasion resistance of the cured film [17, 18].

In order to avoid undesired loss of transparency or gloss the silica particles need to be ground finely into the formulation. When using structure-modified pyrogenic silicas the influence on rheology can be minimized [19]. A surface treatment with e.g., methacryl silane allows the silica particle to be chemically bonded to the film during cure, further improving the mechanical resistance properties of the cured film (Figure 169).

5.7.3 Radiation curing paints for professional painters

Even for non industrial applications radiation curing may become a valid approach to save time and cost when applying coatings for maintenance purposes in or at buildings. Major coatings producers and raw material suppliers work on the subject of decorative coatings cured by a combination of infrared and UV A irradiation [20]. There are still hurdles to overcome, but with the development of tailor-made curing equipment and some raw material optimizations it appears probable that professional painters can make use of this technology soon.

5.8 Outlook

There are many more applications that currently use radiation curing technology on a standard basis which could not be discussed in this book. And there are even more to come in the future. The benefits of this solvent-free technology gain special importance when it comes to the discussion of global warming and the necessary reduction of CO_2 emissions. Besides this the enormous increase in productivity and the possible end-use properties achieved when using radiation curing formulations will more and more make up for the higher raw-material cost and ensure dynamic future growth of this technology.

References

[1] S. Santandrea, Radtech Conference Barcelona, 2005
[2] M. Rüter, Radtech Conference Barcelona, 2005
[3] The Printing Ink Manual, edited by R.H. Leach, Fourth Edition, 1989, p. 11
[4] The Printing Ink Manual, edited by R.H. Leach, Fourth Edition, 1989, p. 20
[5] M. Walkling, Radtech Conference Barcelona, 2005
[6] P. Saunders, SunJet, UV Ink Jet Printing and Ink Technology for the Digital World, 2005
[7] O. Starzmann, Radtech Conference Barcelona, 2005
[8] K. Dietliker, First Commercial Phtolatent Base Catalyst for UV-A Clearcoat Applications, 8th Nürnberg Congress, 2005
[9] M. Kutschera, BASF, Scratch resistance of automobile clearcoats, Journal of Coatings Technology and Research, Volume 3, Number 2 / April 2006
[10] N. Dogan, Akzo Nobel Coatings, A New Generation of UV-A Curable Clearcoats, 8th Nürnberg Congress, 2005
[11] G. Cohen, Status of UV&EB in North America, Radtech Status Report 8_06
[12] Degussa 4 radiation-curing Coatings & Inks
[13] G. Cohen, Status of UV&EB in North America, Radtech Status Report 8_06
[14] P. C. Bouten, Philips, Optical Fiber Coatings: High Modulus Coatings for Fibers With a Low Microbending Sensitivity, Polymer Engineering and Science, Volume 29, No. 17, 1989
[15] AWA Release liner Conference, Chicago 2007
[16] S. Oestreich, Radtech Conference Basel, 2001
[17] G. Leder, Progress in Organic Coatings 45 (2002) 139-144
[18] M. Biehler, Radtech Conference Basel, 2001
[19] S. Frahn, Degussa AG, New modified silicon dioxides for radiation-cured coatings, Aerosil publication, 2002
[20] B. van de Velde, Radtech Congress, Barcelona, 2005

6 Special test methods

All test methods applied in conventionally curable formulations are used also for investigating radiation curable coatings and printing inks. The choice of method is determined by the required property profile, which in turn depends on the intended use. These methods are generally described in reference books and standards. Noted below are brief descriptions of tests that are particularly relevant for raw materials, formulations of radiation-curable systems, and RC cured films.

6.1 Characterization of raw materials and uncured formulations

6.1.1 Chemical functionality

Tests for determining chemical functionality are applied mainly to characterize monomers, oligomers, and additives to measure the content of functional groups and verify the quality of the raw materials.

6.1.1.1 Acid number

The acid number of a raw material is according to DIN EN ISO 3682 the quantity in milligrams of potassium hydroxide (KOH) required for neutralization of 1 g of the substance. For example, an oligomer with an acid number of 56.1 mg KOH/g, contains 1 mmol of acid groups per gram. Determination of a raw material's acid number is important for the following reasons:

- acid groups may have a positive effect on the adhesion of films to metals
- acid groups can react with epoxy resins of cationic systems under certain conditions.
- high acid number can significantly reduce resistance to water and basic components (such as household chemicals)
- acrylate functional monomers and oligomers prepared by esterification with (meth) acrylic acid may have an acid number that depends on the preparation conditions (see Chapter 3.3.3). The acid number indicates the content of unreacted acid, and should lie below 1 mg KOH/g (for reasons of odor and occupational hygiene).

6.1.1.2 Hydroxyl number (OH number)

The OH number is determined by an indirect titration method. According to DIN EN ISO 4629, the OH groups are first acetylated with an excess of acetic anhydride (Figure 170). Excess anhydride is saponified and, along with the acetic acid formed, neutralized with KOH. If the raw material contains additional acid groups, these are also detected by the back-titration so that the measured result is too low and the acid number must be added to the measured value. The OH number of a raw material indicates the amount in milligrams of KOH that is equivalent to the quantity of acetic acid reacting with 1 g of the substance during acetylation.

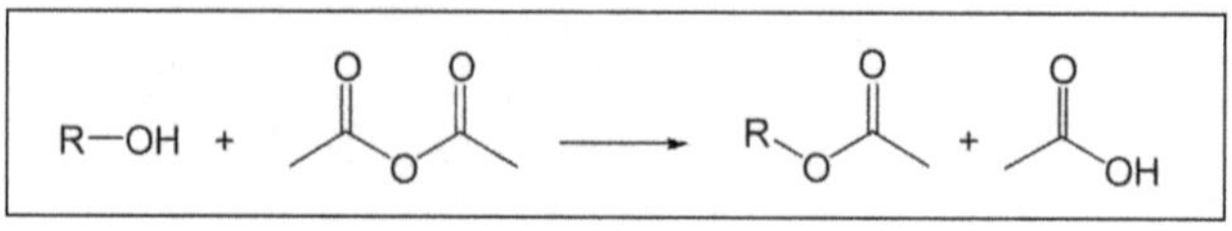

Figure 170: Acetylation of OH groups

Glöckner/Jung/Struck/Studer: Radiation Curing for Coatings and Printing Inks

ISBN: 978-3-86630-907-4

A quantification of the OH number of a raw material is important for the following reasons:

- monomers and oligomers with an OH number carry non-acrylated end groups. If, for example, TMPTA has an OH number of 230 mg KOH/g, only two of the three OH groups are acrylated. The use of this raw material would therefore result in a lower crosslink density than is intended
- OH groups function as chain-transfer agents in cationic systems
- in dual cure systems, OH groups are required for crosslinking with isocyanates

6.1.1.3 Iodine number

The iodine number is a measure of the degree of unsaturation of a sample (DIN 53241-1). It indicates the amount of halogen, in grams of iodine, absorbed by 100 g of a sample under defined conditions. The measured result might possibly be falsified by conjugation of C=C to C=O double bonds. This method is therefore not recommended for acrylates, which should be characterized by, for example, IR or ^{1}H NMR spectra.

6.1.1.4 Epoxy value/epoxy equivalent weight

The epoxy value is the number of moles of epoxy groups per 100 g of the substance (DIN 53188). The epoxide equivalent weight is also frequently encountered (DIN EN ISO 3001); this is the number of grams of the substance required to give one mole of epoxy groups. It is important to know the epoxide functionality of cationically curable substances.

6.1.2 Rheological tests for coatings and printing inks

6.1.2.1 Theoretical principles

Rheology is concerned with investigating the flow behavior of polymer solutions (clear coatings) and suspensions (pigmented systems) [1, 2]. Characterization of the flow behavior of uncured coatings and printing inks is of fundamental importance along the entire process chain.

<u>Preparation of the formulation</u>

A low viscosity while formulating facilitates homogenization of the components. However, if the viscosity during the dispersion of the pigments is too low (e.g., as a result of very high shear thinning), the high shear forces that are necessary to destroy pigment agglomerates cannot be applied. The wetting of the pigments is then not optimal, which can result in sedimentation problems or color aberrations.

<u>Storage and transport</u>

Inorganic pigments of high density in particular tend toward sedimentation. During storage (no shearing action) and transport (very low shearing), therefore, the viscosity should be as high as possible to prevent sedimentation of the pigments.

<u>Application</u>

Good flow and leveling is desired during application, but the liquid film must not run off. For many types of application, shear thinning is desirable so that a low viscosity during application leads to good flow, the viscosity then increasing on the substrate in order to prevent sagging. Because shear forces differ for the various application methods, the material must be chosen so as to suit the method ($\dot{\gamma}$ = 1 to 1000 s^{-1} for painting, curtain coating, or flow coating; $\dot{\gamma} \geq 10^4$ s^{-1} for spraying). A viscosity range from 100 and 300 mPa·s is ideal for spray application. Painting or roller applications require a viscosity between 3,000 and 6,000 mPa·s.

Curing

The temperature of the film can rise up to 80 °C during irradiation. Because viscosity decreases with increasing temperature, defective areas such as curtains due to sagging, or thinning out of the film at edges, may occur as a result of run-off during crosslinking. The lowering of viscosity may cause coatings and printing inks applied on absorbent substrates to penetrate so deeply into the substrate that they are no longer accessible to the radiation. This can result in unsatisfactory curing (see Chapter 3.3.1).

The viscosity of most of the monomers lies between 3 und 150 mPa·s. Depending on their chemical structures, oligomers may possess viscosities between 50 and 5,000 mPa·s (e.g., polyethers, some polyesters) and up to 100,000 mPa·s (epoxy acrylates, urethane acrylates). The higher the tendency of oligomers to form hydrogen bonds, the more marked is their shear thinning flow behavior. However, the flow behavior of formulations depends strongly also on all the other components present. Additives such as layer silicates, fumed silicas, or special additives (see Chapter 3.5.2) can also be used to control the rheological properties.

6.1.2.2 Testing of flow behavior

A number of different measurement systems exist, ranging from simple flow cups to complex oscillation rheometers. In choosing a measurement system, the shear conditions arising during preparation and processing should be taken into account.

Capillary viscometers

Flow cups represent the simplest and therefore the most economical (capillary) viscometers available. The various types differ in their geometries (DIN, Ford, ISO). A beaker of defined capacity has a hole of defined size in the base. The time required for the coating to run out of the beaker is measured. Because no well-defined flow conditions exist, the method allows only single-point viscosity measurements in relation to other samples. It does not provide information on rheological behavior over a shear range, and is, strictly speaking, applicable only to *Newtonian* liquids. In practice, however, the method suffices also for measuring the viscosity even in slightly shear-thinning systems.

Rotation viscometers

Rotation viscometers allow simple determination of the flow behavior of coatings and printing inks over wide ranges of shear. Various measurement setups are used, depending on the viscosity range and the desired information. A distinction is made between plate/plate, cone/plate, and coaxial cylinder systems, the latter two being most commonly used (Figure 171). The advantages of the cone/plate geometry are the ease of cleaning even after the liquid has hardened, the small quantity of sample required, and the possibility of testing samples at high shear rates. Coaxial cylinder systems are essential for low-viscosity liquids, those possessing yield points, or those containing particles (such as dispersions). Benefits of coaxial systems include lower frictional wear, precise temperature control, and generally more accurate measurements.

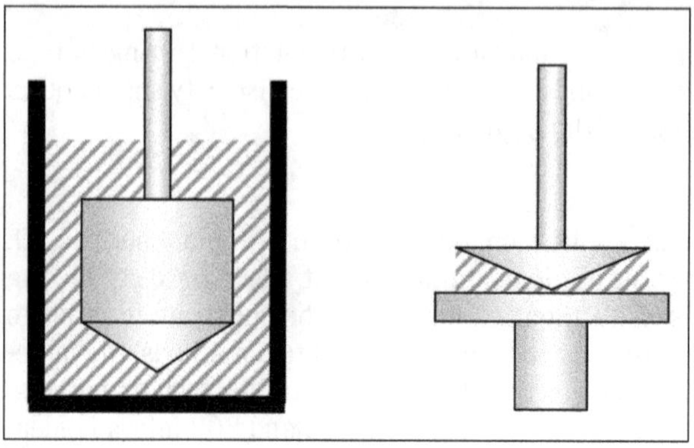

Figure 171: Cylindrical coaxial system (left) and cone/plate measurement system (right)

The different measurement setups yield results that are equivalent only in theory. In any consideration of absolute viscosities, therefore, the measurement system used has to be indicated as well as the shear rate. In practice, these simple measurements are often adequate. In special cases, such as disturbances in flow or wetting of coatings and printing inks, it may be useful to investigate the viscoelastic properties using oscillation rheometers. The underlying theory and the measurement setup are described in [1]. Special rheometers equipped with an irradiation unit also allow monitoring of radiation-induced crosslinking as a function of time [3].

6.2 Testing of the curing process and the cured film

6.2.1 Simple methods for testing cure

The film should generally be resistant to external influences after curing. In practice, surface cure and through-cure are checked using the following easy-to-use methods.

6.2.1.1 Fingernail scratch test

The film surface is examined for damage after drawing the nail of the index finger over the cured film. Slip and surface cure in particular can be judged by this method. The test is highly subjective, as it depends on the force with which the fingernail is drawn over the surface.

6.2.1.2 Thumb twist test

In this test method, the thumb is pressed firmly on the cured film and twisted. The film is examined for damage, such as deformation, and its tackiness is assessed. A soiled thumb indicates an imperfect cure. This test is again highly subjective.

6.2.1.3 Powder test

Talc, carbon black, or some other powder can be sprinkled on to the surface of the cured film to assess surface tack. Any powder not adhering to the surface is removed by compressed air or a soft brush. The more powder that adheres to the film, the higher is the tack. Cotton swab, for example, can be used as an alternative by sliding it over the surface and observing fibers adhere to the film.

6.2.1.4 Adhesive tape test for through-cure

The adhesive tape test is usually performed in order to check the quality of the adhesion of printings inks. Furthermore, the degree of through-cure of the film can easily be checked by applying the formulation on a release paper and curing under the selected conditions. The film is then drawn off from the substrate by means of adhesive tape. If the film appears greasy and matt on the side facing the substrate, through-cure has not occurred, and optimization is required e.g., to improve adhesion and to reduce the content of extractable components.

6.2.1.5 Solvent test (MEK or acetone test)

The solvent test can be used to assess whether the crosslink density is sufficiently high. A swab saturated with a solvent is moved up and down the film surface by means of a 1 kg hammer. The number of to-and-fro cycles is counted. After each cycle, ones checks whether the film surface is swelled, partially dissolved, or has become detached from the substrate. If the film remains undamaged after 100 cycles, resistance to the solvent is satisfactory. In some areas of application, resistance to 150 cycles is required. The most commonly used solvent is methyl

ethyl ketone (2-butanone). Acetone, xylene, isopropanol and butyl acetate are less frequently used. The results of the measurement are influenced by parameters such as surface roughness and thickness of the film, and contact pressure of the hammer during the measurement.

6.2.2 Checking of cure by instrumental methods

6.2.2.1 UV/visible spectroscopy

Most components of RC formulations absorb radiation of shorter wavelength than visible light, and are therefore colorless. Ideally, monomers, oligomers, and additives should absorb only at wavelengths shorter than the emission spectra of radiation sources used for curing, so that largest part of radiation is available for cleavage of the photoinitiators. Photoinitiators, on the other hand, must absorb within the spectrum emitted by the radiation source in order to be activated (see Chapter 3.2).

UV and visible absorption spectra of raw materials and formulations can be recorded in strongly diluted form in a UV/vis spectrometer. The measurement principle is described in [4]. Figure 172 shows the UV/vis spectrum of a photoinitiator that absorbs particularly strongly in the region between 250 and 280 nm. It can be checked in this way whether the absorption spectrum of the photoinitiator used is compatible with the emission spectrum of the radiation source used for curing (see Chapter 4.1.1).

Examples of application

- In most cases, the decomposed photoinitiator absorbs in a different region than its initial form. The radiation dose required for complete decomposition can be determined by observing the reduction of photoinitiator concentration during irradiation as a function of time, or radiation intensity.
- Recording of UV/vis spectra of formulations provides information on whether absorption by formulation components (such as UV stabilizers) overlaps the absorption maximum of the photoinitiator, thus negatively impacting the cure.
- A UV/vis spectrum of the cured sample may be recorded to check the concentration of unreacted photoinitiator in a coating film. This allows optimization of the concentration of photoinitiator in the formulations.

6.2.2.2 Photo differential scanning calorimetry

Conventional DSC equipment can be used to monitor phase transitions such as melting points, glass transition temperatures, and reaction enthalpies. For investigation of the curing process in radiation curable systems, the device can be equipped with an irradiation unit. A photo differential scanning calorimeter (photo DSC) of this type allows the crosslinking reaction to be monitored as a function of irradiation time under various conditions. In this way, the composition of the formulation (e.g., the type and concentration of the photoinitiators) can be coordina-

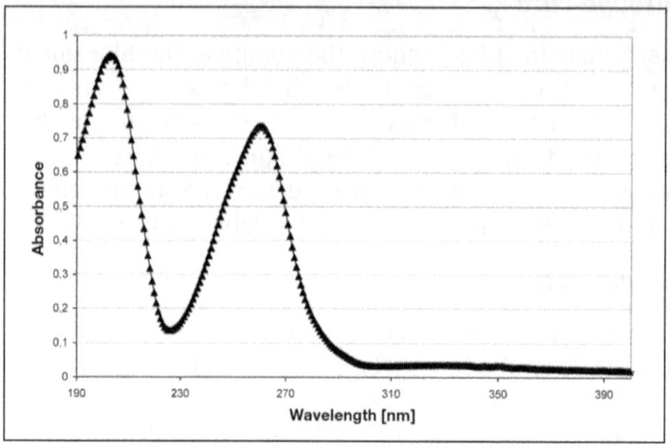

Figure 172: UV/vis spectrum of a photoinitiator (0.01 % "Irgacure" 127 in acetonitrile, film thickness 0.1 cm)

ted with various process parameters such as type and intensity of the radiation source, line speed, and curing temperature, so as to optimize the efficiency of the curing process.

The test is performed by irradiating the sample with UV light in the measuring device. The heat generated during polymerization is recorded as for a conventional DSC. The heat generated by the UV lamp is determined by a parallel measurement on a sample already polymerized, and is subtracted from the measurement signal. The measurement gives the heat flow arising from the polymerization as a function of time (Figure 173). The area over the heat-flow curve corresponds to the heat of reaction, which increases with the number of functional groups that react. This allows investigation of the influence of different curing parameters (e.g., temperature, radiation intensity, radiation sources with various emission spectra). The subject is discussed in more detail in [5, 6].

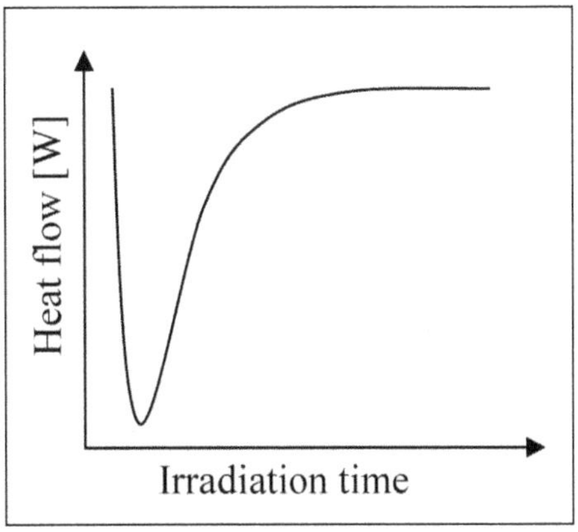

Figure 173: Heat flow curve for irradiation of a sample as a function of time

6.2.2.3 Fourier transform infrared (FTIR) spectroscopy

Oligomers have an IR absorption spectrum, each peak characterizing a chemical bond. In IR spectroscopy, bond energies are expressed in wavenumbers, [cm^{-1}], corresponding to the reciprocals of the wavelengths. Some functional groups and their characteristic wavenumbers are shown in Table 41.

The cure efficiency of a UV curing process can be monitored by transmission FTIR spectroscopy. The UV curable formulation is applied on a NaCl, BaF_2, CaF_2, or KBr crystal by means of a calibrated wirewound bar coater (with DFT typically 10 to 20 µm), and the infrared spectrum of the clear coat is recorded prior to and after UV exposure. The polymerization is monitored by following the decay of the IR bands (Table 41). The degree of conversion (x) is directly related to the decrease of the IR absorbance. It can be calculated from the following equation:

$$Conversion\ [\%]\ x = \frac{A_0 - A_1}{A_0} \cdot 100$$

where A_0 and A_t represent the area of the IR band before and after UV exposure over time t.

The conversion at the film surface can be evaluated by means of attenuated total reflectance (ATR) FTIR spectroscopy. The sample is applied directly on the diamond of an ATR unit. An evanescent IR beam penetrates the surface of the film in contact with the diamond, generating the surface FTIR spectrum of the UV cured coating (Figure 174). By removing the film from the substrate, this tool also allows measurement of the conversion at the bottom layer in contact with the substrate, thus characterizing the through-cure of a cured coating.

Table 41: Characteristic absorptions of selected functional groups [4]

Reactive group	Wave number [cm^{-1}]
acrylate	795 to 815 (~1635; ~1410)
epoxide	800 to 910
vinyl ether	920 to 975
isocyanate	2225 to 2275

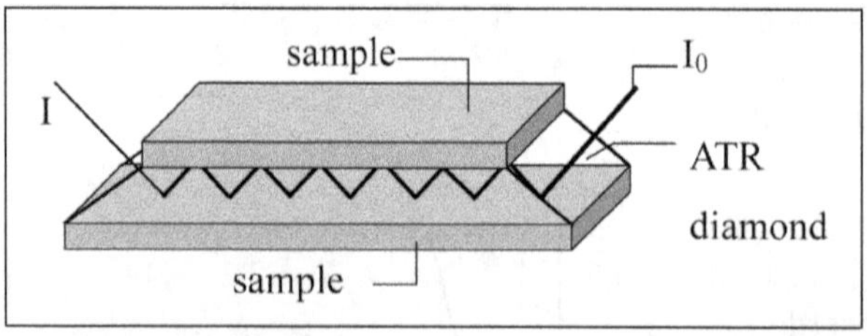

Figure 174: Principle of the ATR technique; I_0, I = IR intensity before and after absorption

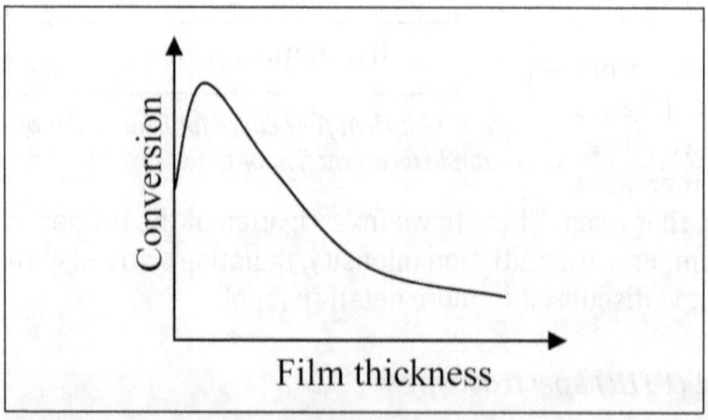

Figure 175: Degree of conversion as a function of film thickness in a radically curable film

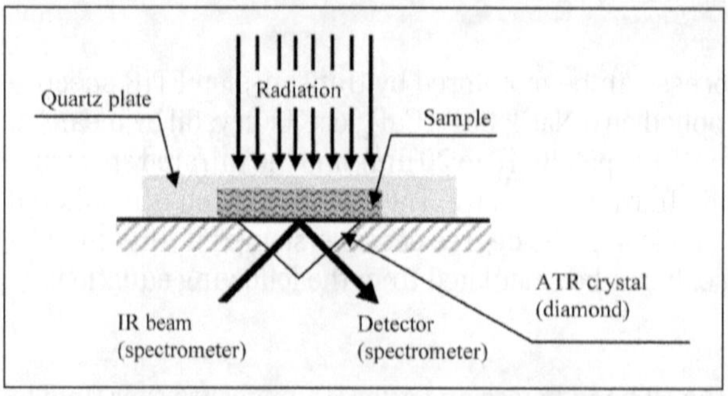

Figure 176: Measurement setup for investigation of radiation-induced polymerization by RT ATR FTIR [7]

This equipment can also be used for pigmented films.

For detailed information on conversion of the polymerizable groups as a function of film thickness, thin layers of the sample film can be prepared (by slicing with a microtome). After recording the ATR FTIR spectra, the degree of conversion is obtained as a function of film thickness. Figure 175 shows the example of a radically curable system in which the degree of conversion directly at the surface is relatively low, as a result of oxygen inhibition. After reaching a maximum, the degree of conversion decreases as a result of the inner filter effect which diminishes light intensity within the film in accordance with the *Lambert-Beer* Law.

Real-time (RT) FTIR spectroscopy (ATR or transmission) is a powerful tool for comparing cure speed and efficiency of different formulations. A light source, usually a medium-pressure mercury lamp equipped with an optical guide, is used to induce polymerization. FTIR spectra are simultaneously recorded, thus allowing the polymerization on UV exposure to be monitored in real time (Figure 176). Here again, the degree of conversion (x) is directly related to the decrease of IR absorbance, and can be calculated from the above equation. From the conversion versus time curves recorded by RT FTIR spectroscopy (Figure 177), it is possible to evaluate the maximum rate of polymerization $(R_p)_{max}$ from the product of the maximum slope $(dx/dt)_{max}$ (also called reactivity) and the initial concentration $[M]_0$ of, e.g., acrylate double bonds:

$$\left(R_p\right)_{max} = \left(\frac{dx}{dt}\right)_{max} [M]_0 .$$

6.2.2.4 NIR spectroscopy

Transmission IR spectra of thick films (above 50 μm) do not allow quantitative evaluation of the extent of cure, due to saturation of the IR absorption. Modern spectrometers also measure in the near-IR region [8]. The degree of cure of films having thicknesses of up to several mm can be monitored via the acrylate overtone bands at 6150 to 6175 cm^{-1} and 4730 to 4740 cm^{-1}.

6.2.2.5 *Volume shrinkage*

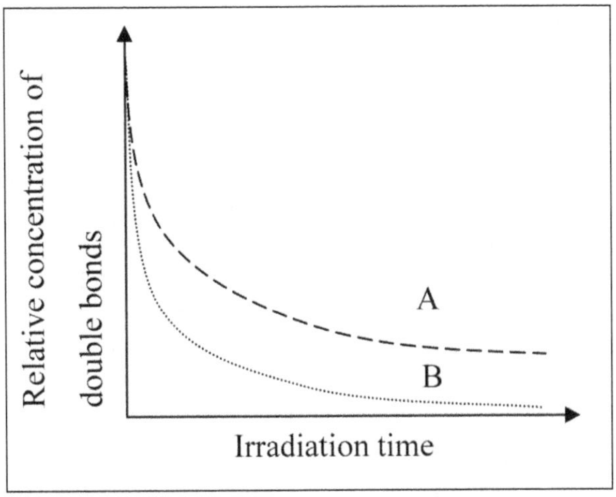

Figure 177: Time dependence of the conversion of acrylate double bonds by irradiation. Sample B shows faster and more complete polymerization than Sample A

As seen from Table 9, adhesion of radiation curable formulations is influenced by several parameters. If all these parameters have been optimized and adhesion needs to be further improved, it may be necessary to reduce the internal stress caused by volume shrinkage during UV cure (see Chapter 3.3.1.4). Methods for determining internal stress are not available to every developer; these include, for example, the cantilever beam method, described in detail in [9]. Density determination, on the other hand, is possible in most laboratories and, on account of the proportionality between internal stress and volume shrinkage, provides information on the possible cause of the adhesion problems.

Because density increases as a result of volume shrinkage during polymerization (see Chapter 3.3.1.4), the relative volume shrinkage is easily obtained by determining the density of the formulation before (d_M) and after (d_P) curing.

$$\text{Shrinkage [\%]} \quad \Delta V = \frac{d_p - d_M}{d_p} \cdot 100$$

The density of the liquid formulation can be measured using, e.g., a liquid displacement pyknometer, by determining the mass of a defined volume. The density of the cured film can also be determined pyknometrically in the following way. The liquid formulation is applied to, e.g., a release paper. The film is then cured, the mass of the free film determined, and its volume measured pyknometrically (using a gas or liquid displacement pyknometer) [10]. The cure has to be performed under conditions close to those in actual practice, because a difference in the degree of monomer conversion strongly influences the measured values. A detailed overview of static and dynamic methods for determining volume shrinkage is available in [9].

6.2.3 *Determination of migratable components*

The migratable components in a formulation and subsequent coating film are not incorporated into the film during curing and can be dissolved out on account of their solubility. These can include, in addition to raw materials, unreacted decomposition products of the photoinitiators [11].

Migratable components can accumulate at the surface of the film by migration. From there, they may migrate from the package into the foods themselves, or reach the skin via direct contact. During production of a packaging material, the printed exterior side of the packaging might also be in contact with the interior which will further contain the food, thus depositing migrating species; this might occur by, e.g., rolling up of freshly printed sheeting. A recent example is the detection of the photoinitiator ITX (isopropyl thioxanthone) in fruit juices and baby foods. Following these findings, the use of ITX has been heavily restricted.

Materials, such as food packaging, that are intended to come into contact with foodstuffs must be produced so that no component could migrate to the food in such quantities that would pose a health risk or affect the taste or odor of the food [12, 13]. However, assessment of the health risks of coatings and printing inks in contact with foods is not regulated at EU level so far. An EU harmonized regulation (Directive 2002/72/EC, relating to plastic materials and articles intended to come into contact with foodstuffs) is currently available only for plastics [14]. This can apply only if the coatings and printing inks are composed of the same monomer building blocks, or contain the same additives. According to the directive, the only authorized starting substances which can be employed for coating the inside of packaging (the side in contact with the food) are regulated in the positive list with restrictions and limit values [14]. For the outside of the packaging, substances that are not listed may also be used if they do not migrate into the food, i.e., if they are not detectable in the food (detection limit: 10 ppb). The total of all migratable substances (overall migration, global migration) must not exceed a limit of 60 mg/kg. Specific migration limits (SML) have been defined for individual substances.

To quantitatively determine migratable components, the formulation is applied to a substrate of a defined area and cured. The film is then immersed in a food simulating solvent at a defined temperature and over a specified period of time (in the EU, e.g., 40 °C over 10 days). The dissolved substances out of the film are identified and quantitatively analyzed by high pressure liquid chromatography (HPLC), possibly coupled with mass spectrometry (HPLC-MS). Various simulant solvents represent the different classes of foods. For aqueous foods with a pH value above 4.5, distilled water is used, and for a pH below 4.5, 3 % acetic acid. Alcoholic foods are simulated by 10 % ethanol, and fatty foods by olive oil. Because olive oil is not a single pure substance, isooctane, 95 % ethanol, or the triglyceride HB 307 can be used as alternatives to simplify testing [13]. Gas-phase migration is simulated by means of a Tenax (polyphenylene oxide) column.

Specific migration of unlisted substances (from the outer layer, not the direct contact layer) of less than 10 ppb currently requires no further toxicological investigation within the EU. If higher concentrations are detected, toxicological investigations are required as follows: 10 to 50 ppb: Ames test, chromosomal aberration test, mouse lymphoma test; 50 to 5000 ppb: additionally, 90 day feeding study in rats; >5000 ppb: 2 year feeding study in rats (carcinogenicity study).

References

[1] M. Pahl, W. Gleißle, H.-M. Laun, "Praktische Rheologie der Kunststoffe und Elastomere", 4. überarb. Auflage, VDI-Verlag, Düsseldorf, Germany (1995)

[2] T. Brock, M. Groteklaes, P. Mischke, Hrsg. Ulrich Zorll, Lehrbuch der Lacktechnologie, 2. berichtigte Auflage, Vincentz Verlag, Hannover, Germany, 2000

[3] C. Gonzalez, UV Curing Characterization using Rheology: Coatings, www.rheologysolutions.com

[4] M. Hesse, H. Meier, B. Zeeh, Spektroskopische Methoden in der organischen Chemie, 4. überarbeitete Auflage, Georg Thieme Verlag, Stuttgart, New York, 1991

[5] E.W. Nelson, J.L. Jacobs, A.B. Scranton, Polymer, 36 (24), 1995, p. 4651 - 4656

[6] H. Thiem, P. Strohriegl, M. Shkunov, I. McCulloch, Macromolecular Chemistry and Physics, 206, 2005, p. 2153 - 2159

[7] C. Elsner, M. Buchmeiser, Nachrichten aus der Chemie, 54, 2006, p. 141 - 144

[8] T. Scherzer, RadTech Europe 2007, Conference Proceedings

[9] J. Stansbury, Junhao Ge, RadTech Report, May/June 2003, p. 56 - 62

[10] J.M. Blanding, C.L. Osborn, S.L. Watson, Journal of Radiation Curing, 4, 1978, p. 13 - 28

[11] J.R. Veraart, RadTech Europe 2007, Conference Proceedings

[12] European Parliament, Regulation (EC) No. 1935/2004, 2004

[13] T. Simat, Nachrichten aus der Chemie, 54, September 2006, 2006

[14] European Commission, Directive 2002/72/EC, 2002

7 Environmental and occupational protection

Radiation curable coatings and printing inks have a low VOC content because they contain no volatile organic solvents. Emissions in surrounding environment are therefore low, and post-combustion and recycling plants usually not required. The flash point is usually above 100 °C. Recycling of, e.g., overspray is relatively easy. Energy requirements for curing are low as compared with those for conventional thermosetting systems. All in all, the use of radiation curable coatings and printing inks reduces environmental impact. However, the starting materials for RC compositions may pose a certain risk to human health and the environment. The following protective measures are among those that must be observed when handling these materials:

- obtain information on possible risks (e.g. MSDS) before starting the work
- ensure adequate ventilation and extraction facilities in the workplace; avoid inhalation of vapors
- avoid direct contact with components and formulations
  - use personal protective equipment (e.g., safety spectacles, protective clothing, gloves)
  - in case of contact, remove contaminated clothing immediately and wash the contaminated area; in case of contact with eyes or mucous membranes wash thoroughly and consult a physician
- do not allow the substances to escape into the atmosphere

The risk potential of the starting materials depends in each case on the class of raw materials [1]. Most raw materials are only slightly volatile at room temperature, so that the risk of damage to respiratory organs by inhalation is very slight if the materials are correctly handled.

Inhalation of or contact with acrylate functional monomers and oligomers can cause primarily irritation of the respiratory tract, skin, and mucous membranes. Epoxides and especially vinyl ethers usually have lower irritancy than acrylates.

For all substances, risk of skin sensitization exists by direct contact. As a general rule, the risk potential increases with decreasing molecular weight, so that monofunctional monomers have a higher risk potential than, e.g., ethoxylated acrylate monomers or oligomers (Table 42). Certain raw materials have tested positive in animal trials, or are suspected of causing cancer. Examples include diethoxyacetophenone, n-vinyl pyrollidone, and neopentyl glycol diacrylate.

Possible symptoms during UV curing include burning of the skin due to UV radiation, the severity of which depends on the wavelength emitted and the period of exposure. Eye damage (ranging from inflammation of the eyes to temporary blindness) may additionally be caused by direct irradiation; this is normally reversible if exposure times are short. However, long-term exposure to UV radiation can cause macular degeneration what leads to definitive loss of central vision. UV light below 260 nm can result in the formation of ozone, which if inhaled can lead to headache, fatigue, and irritation of the eyes and lungs;

Glöckner/Jung/Struck/Studer: Radiation Curing for Coatings and Printing Inks

ISBN: 978-3-86630-907-4

Table 42: Examples of monomers and their risk potential [2]

Monomer	Hazard symbol	R-phrase no.	R-phrase text
IBOA	Xi, N	R36/37/38 R51/53	irritating to eyes, respiratory system, and skin Toxic to aquatic organisms, may cause long-term adverse effects in the aquatic environment
HDDA	Xi	R36/38 R43	irritating to eyes and skin may cause sensitization by skin contact
DPGDA	Xi	R38 R41 R43	irritating to skin risk of serious damage to eyes may cause sensitization by skin contact
TPGDA	Xi, N	R36/38 R43 R51/53	irritating to eyes and skin may cause sensitization by skin contact toxic to aquatic organisms, may cause long-term adverse effects in the aquatic environment
TMPTA	Xi	R36/38 R43	irritating to eyes and skin may cause sensitization by skin contact
TMP(EO)3TA	none	none	none

radiation below 260 nm can easily be absorbed by quartz filters. The energy of non-ionizing UV radiation is too low to penetrate the body. High-energy electron beams on the other hand can penetrate the body and destroy cells, damaging the blood and central nervous system. For this reason, only professional equipment with high safety standards should be used, which should be operated only by trained staff. All contact with radiation must be strictly avoided.

It must also be borne in mind that certain raw materials can form toxic substances on irradiation. Some photoinitiators that contain sulfonium salts can release benzene.

The cured films generally have no hazard potential. However, incomplete conversion may cause small quantities of low-molecular products to remain in the film, which can then migrate toward the film surface. For this reason, manufacturers of raw materials offer products with a low tendency to migration, such as polymeric photoinitiators and alkoxylated monomers.

References

[1] K. Lawson, C. Robinson, in J.P. Fouassier, J.F. Rabek, Radiation Curing in Polymer Science and Technology - Vol. IV, Elsevier Applied Science, London, New York, 1993

[2] see MSDS of monomer supplier

8 Trouble shooting

The following tables might help the technician to solve routine problems arising during the manufacture and processing of radiation curable systems.

8.1 Properties of the liquid coating/ink

Issue	Possible causes	Possible solutions
1. Poor storage stability	Formation of radicals over time, due to insufficient stabilization (e.g., increase in viscosity, gelation)	Increase the proportion of stabilizers to 50 to 300 ppm. The concentration should be as low as possible to avoid curing and coloration problems (see Chapter 3.5.5).
	Formation of radicals by exposure to light (e.g., increase in viscosity, gelation)	Storage under UV/light-proof conditions.
	Formation of hydroperoxide radicals by interaction with atmospheric oxygen (e.g., increase in viscosity, gelation)	Storage under inert conditions.
	Formation of radicals by exposure to heat (e.g., increase in viscosity, gelation)	Protect from temperatures above 60 °C; add stabilizers.
	Shear forces (e.g. pumping)	Avoid high shear rates if possible. Stabilize pigments and fillers as thoroughly as possible (e.g., by use of pigment wetting and dispersing additives).
	Decrease of reactivity by re-crystallization of photoinitiator	Heating and stirring. Use of liquid or amorphous photoinitiators.
	Sedimentation of pigments and fillers	Add pigment wetting and dispersing additive. Adjust rheology to slightly thixotropic flow behavior if possible (e.g., by the use of silica).
2. Rheology problems	Viscosity too high (also under shear)	Choose lower viscosity oligomers. If film properties allow, increase monomer content. Switch to monomers with higher solvating power (see Chapter 3.3.3). Addition of rheology modifiers.

Table continued on next page

Glöckner/Jung/Struck/Studer: Radiation Curing for Coatings and Printing Inks

ISBN: 978-3-86630-907-4

Issue	Possible causes	Possible solutions
2. Rheology problems	Viscosity too high (but decreases under shear)	Check oligomers for their potential to influence shear thinning flow. Use suitable pigment wetting and dispersing additives. Reduce amount of small particle size ingredients (pigment, filler, matting agent,...) if possible.
	Sagging/running of coating applied to vertical surfaces	Create shear thinning flow behavior by using thickening agents (see Chapter 3.5.2). Reduce film thickness if possible.
3. Foam problems	Raw materials with foam stabilizing properties (often already apparent during manufacture and application)	Select raw materials (especially surfactants) with low foam stabilization. Add defoamer/deaearator (see Chapter 3.5.4).
	Insufficient deaeration (apparent mainly in applied film)	Add deaerator (see Chapter 3.5.4). Select raw materials (especially surfactants) with low foam stabilization.

8.2 Curing behavior and mechanical properties of the cured coating/ink

Issue	Possible causes	Possible solutions
4. Liquid or poorly cured film after irradiation	Insufficient cure due to oxygen inhibition, see Chapter 2.1.1.4 (free radical cure only)	Increase radiation intensity. Increase radiation dose. Increase photoinitiator concentrations. If yellowing is acceptable, change photoinitiator to low oxygen sensitive α-aminoketone. Introduce amine synergist (see Chapter 3.3.2.5 and 3.2.2.3). Use deaerators. Irradiate under an inert atmosphere (e.g., nitrogen blanketing). In thin layers: increase viscosity and/or reduce film temperature. In thick layers: decrease viscosity and/or increase film temperature, and work under inert conditions.
	Insufficient cure due to: moisture inhibition base inhibition, see Chapter 2.1.2 and 3.2.3 (cationic cure)	Reduce/check ambient humidity. Avoid basic formulation components and substrates.
	Low photoinitiator concentration: curing not adequately initiated (see Chapter 3.2)	Increase photoinitiator concentration.

Table continued on next page

Issue	Possible causes	Possible solutions
4. Liquid or poorly cured film after irradiation	Low light intensity and/or dose: curing not adequately initiated (see Chapter 4)	Increase lamp power. Increase irradiation time (e.g., by slowing down conveyor belt). Clean lamp and reflectors.
	Poor overlap of photoinitiator absorption and lamp emission poor match of absorption curves of photoinitiator, pigment and UV absorbers, see Chapter 2.1.1.1 (UV cure only)	Check absorption and emission curves, and adapt photoinitiator type to lamp emission accordingly; if required add sensitizer. Adapt photoinitiator type to pigment and UV absorbers.
	Poor reactivity of oligomers or monomers	Select more reactive oligomers and/or monomers (of higher functionality). Avoid unsaturated polyesters, methacrylates, use epoxy acrylates. • Use up to 15 % of vinyl ethers. • See also Chapter 3.3.1.3.
5. Tacky surface	Refer to point 4	Refer to point 4.
	Photoinitiator not suited for surface cure (UV cure only)	Add surface-active photoinitiators (hydroxyketones or phenylglyoxylates).
	Poor emission at short wavelengths (UV cure only)	Check lamp emission spectra and switch to medium pressure mercury lamps with strong emission in UV-C and UV-B regions.
6. Poor through-cure	Inner filter effect (see Chapter 2.1.1.4)	Reduce film thickness to the minimum. Increase radiation dose. Add acylphosphine oxides or increase their concentration. Increase UV-A emission (e.g., Ga-doped Hg-lamp, fluorescent lamps). Pigmented films: adjust transmission window of pigments with absorption of photoinitiators.
7. No cure in shadow areas	Light cannot reach shadow areas (UV cure only)	Use dual-cure systems. Illuminate as well as possible by the use of reflectors (see Chapter 4.1.2.4) "Larolux" "PlasmaCure"
8. Insufficient overall hardness	Low crosslink density due to insufficient cure	Refer to points 4 to 6. Increase curing temperature.
	Low crosslink density due to raw materials	Increase content of highly functionalized monomers and oligomers (see Chapter 3.3.1.3). Change monomer/oligomer ratio in favor of monomer. If polymerization inhibitor present, reduce its concentration.

Table continued on next page

Issue	Possible causes	Possible solutions
8. Insufficient overall hardness	Low glass transition temperature	Increase crosslink density (see above). Change from linear aliphatic to cycloaliphatic and aromatic raw materials. Add oligomers with a high density of hydrogen bonds (e.g., epoxy acrylates, urethane acrylates). Add high Tg non-reactive resins (e.g., ketone-aldehyde resins).
	Any other cause	Add suitable fillers (e.g., silica) that increase the overall hardness of the film.
9. Poor scratch resistance	Insufficient cure/hardness	Refer to points 4 to 8. Introduce nanoparticles like silica (see Chapter 3.5.7 and 3.6.6).
	Blunt surface	Introduce slip additive (e.g., waxes or polyether siloxanes, see Chapter 3.5.6.1).
	Uneven surface	Improve flow and leveling (e.g., by reducing viscosity through addition of monomers or organic solvents, or by the use of suitable flow additives).
10. Poor blocking resistance	Insufficient cure	Refer to points 4 to 8.
	Excessively low glass transition temperature of resulting polymer film	Refer to point 8.
	Any other cause	Introduce additives that improve blocking resistance, such as waxes, siloxane-based additives and combinations of both (see Chapter 3.5.6.1. and 3.5.7) Add suitable fillers (e.g., silica) that increase the overall hardness of the film.
11. Poor abrasion resistance	Insufficient cure	Refer to points 4 to 8.
	Non-optimal glass transition temperature (too high or too low) of resulting polymer film	Choose the most suitable oligomers and monomers (e.g., polyurethane acrylate, if necessary in combination with polyether acrylates).
	Any other cause	Add suitable fillers (e.g., silica) that increase the overall hardness of the film.

Table continued on next page

Issue	Possible causes	Possible solutions
12. Controlled coefficient of friction (COF) difficult to obtain	High amounts of migratory components after cure additive package not ideal	Ensure good and fast through-cure (refer to 6.). Use very strong slip agents above the minimum dosage for low COF surfaces (see Chapter 3.5.6). Use milder slip agents above the minimum dosage for medium COF coatings (instead of merely reducing the amount of strong slip agent). Ideally, use crosslinkable additives (silicone acrylates) (see Chapter 3.3.2.5 and 3.5.6.1). Possibly use both silicone-based glide additives and waxes.
13. Poor resistance properties (chemicals, solvents, food ingredients, water)	Low crosslink density due to insufficient cure	Refer to points 4 to 7.
	Low crosslink density due to raw materials	Increase content of highly functionalized monomers and oligomers. Change monomer/oligomer ratio in favor of high functional monomer. If polymerization inhibitor present, reduce its concentration.
	Non adapted hydrophilic/hydrophobic properties	E.g., in case of poor water resistance: avoid highly hydrophilic formulation components (ethoxylated monomers, polyether acrylates, surfactants, etc.). Choose oligomers and monomers that are as hydrophobic as possible (e.g., monomers with long aliphatic side chains). Add internal hydrophobing agent (e.g., one based on silicone resin). Introduce hydrophobic surface control additive that creates water repellency, e.g. waxes and special siloxane-based additives.
	Poor chemical resistance of individual formulation components	Avoid polyether acrylates, ethoxylated monomers and possibly also polyester acrylates. Use in preference polyurethane acrylates and epoxy acrylates.
14. Poor weather resistance	Refer to 13.	Avoid use of amine-containing photoinitiator. Increase concentration of HALS and UV absorbers. Check the cure efficiency.

Table continued on next page

Issue	Possible causes	Possible solutions
14. Poor weather resistance	Poor UV resistance of formulation components	Avoid aromatic structures (e.g., epoxy acrylates), oxidation promoters (e.g., tertiary amines), and high amounts of polyethers and vinyl ethers. Use aliphatic oligomers and monomers (e.g., urethane acrylates and possibly polyester acrylates in free radical cure, or cycloaliphatic epoxies in cationic cure). Introduce UV absorbers and radical scavengers (e.g., HALS).
	Incompletely photolyzed photoinitiator resulting in blooming, yellowing, and surface brittleness on weathering	Increase light dose during curing. Use hydrophobic and/or high molecular weight photoinitiators. Refer to 18.
15. Tearing of the film during cure, possible delamination	Insufficient elasticity excessively high volume shrinkage and film tension	Introduce "flexible" oligomers (e.g., urethane acrylates of high molecular weight, low functionality and low Tg). Avoid high amounts of highly functionalized monomers and oligomers. Introduce chain transfer agents (free radical cure) or alcohols (cationic cure) for internal plasticization. Introduce some monofunctional monomer for plasticizing. Increase curing temperature slightly. Introduce thermoplastic resins (not functionalized). Introduce fillers to reduce film tension (critically check on transparency of clears).
16. Poor adhesion	Poor film/substrate interaction (adhesive forces)	See also Chapter 3.3.1.4. Choose the most suitable oligomer for the respective substrate (e.g., one containing acid groups for metal substrates). Possibly pretreat substrate. Introduce adhesion resins. Introduce adhesion promoting additives, see Chapter 3.5.8.
	Insufficient elasticity excessively high volume shrinkage and film tension	See above. See also Chapter 3.3.1.4.
	Poor through-cure	See above. See also Chapter 3.3.1.4.
	Insufficient or no penetration into the substrate swelling of the substrate	Reduce viscosity as far as possible. Choose monomer with high solvating power (e.g., DPGDA). Possibly add organic solvents.

Table continued on next page

Issue	Possible causes	Possible solutions
16. Poor adhesion	Poor wetting of the substrate poor flow and leveling of the film	Ensure proper cleaning/pretreatment of the substrate. Introduce suitable additives (e.g., polyether siloxanes, silicone acrylates) that lower the surface tension of the uncured formulation.
	Any other cause	Consider switching from free radical cure to cationic cure.
17. Excessively high level of migratory/ extractable components in cured film	Insufficient cure	Refer to points 4 to 8.
	Choice of photoinitiator	See also Chapter 3.2.
	Excessively high amounts of non crosslinkable low molecular weight additives	Preferably use crosslinkable additives (e.g., silicone acrylates) or additives of very high molecular weight (e.g., special pigment wetting and dispersing additives).

8.3 Appearance of the cured coating/ink

Issue	Possible causes	Possible solutions
18. Undesirable yellowing immediately after curing	Choice of photoinitiator (free radical cure only)	Change photoinitiator to bleachable acyle phosphine oxide.
	High level of unphotolyzed photoinitiator	Increase light dose by increasing exposure time and/or light intensity. In some cases (for very high doses), you may need to do just the opposite.
	Formation of colored by-products during photolysis (free radical cure only)	Slightly increase cure temperature and light dose.
	Presence of oxidation promoters	Avoid tertiary amines and, in general, all components susceptible to reaction with oxygen.
19. Surface defects, craters	Insufficient substrate wetting insufficient flow contamination of the substrate or the coating/ink	Introduce suitable additives that lower the surface tension of the uncured formulation (e.g., polyether siloxanes, silicone acrylates) (substrate wetting additives, glide additives). Ensure proper cleaning or possibly pretreatment of the substrate. Produce coating/ink in clean environment.
20. Poor gloss	In clears: incompatibility of components unsuitable oligomers curing temperature not ideal	Select raw materials that are compatible with each other (clear before and after cure). Choose oligomer with good gloss development. Use ketone-aldehyde resins. If possible increase curing temperature.

Table continued on next page

Issue	Possible causes	Possible solutions
20. Poor gloss	In pigmented coatings/inks: refer to clear (see above) poor pigment dispersion	Refer to clears (see above). Select suitable pigment wetting and dispersing additives. Select modified oligomers with improved pigment wetting properties for grinding. Optimize grinding conditions (time, media, temperature, viscosity).
21. Insufficient matting	Not enough matting agent unsuitable grade of matting agent matting agent dispersed too strongly	Increase amount of matting agent. Select special fumed silica tailor-made for radiation-curing (see Chapter 3.5.7). Avoid excessive shear forces in incorporation of matting agent. If using a wetting and dispersing additive: choose a different grade. Reduce dosage or eliminate completely.
22. Poor color strength, hiding power, or transparency in pigmented coatings/ inks	Poor pigment dispersion poor pigment stabilization	Select suitable pigment wetting and dispersing additive. Select modified oligomers with improved pigment wetting properties for grinding. Use ketone-aldehyde resins. Optimize grinding conditions (time, media, temperature, viscosity).

8.4 Sensorial properties of the cured coating/ink

Issue	Possible causes	Possible solutions
23. Strong odor after exposure	Volatile photoinitiator by-products volatile oligomers poor cure efficiency (UV cure only)	Use high molecular weight photoinitiators with non-volatile by-products. Use non-volatile oligomers and monomers. Improve cure efficiency (refer to points 4 to 8).

Authors

Patrick Glöckner is Director of the Technical Service Coatings of EVONIK Tego Chemie GmbH. He joined EVONIK Degussa in 2001 as Manager in the RD-Department, overseeing the Product Development of Specialty Resins. Born in 1971, he started his professional career with Herberts (today: DuPont Performance Coatings) working as a lab assistant in coatings laboratories. He studied chemistry at the University of Wuppertal. His diploma thesis was awarded the APi-Prize 1998 by the German Chemical Society (GDCh), Division "Paints and Pigments". In 2001 he received his PhD in Polymeric Science from the University of Mainz.

Tunja Jung, Ciba Specialty Chemicals Basel, received her diploma degree in chemistry in 1996 at the Science and Technology Academy in Isny, Germany, after completing her diploma work on UV-curable pigmented coatings. In the same year she joined Ciba Specialty Chemicals developing photoinitiators and photolatent compounds for radiation curing, at the Technology Center, Business Line Coatings. Since May 2008 she is heading the New Ventures Group within the Segment Coating Effects.

Susanne Struck completed her studies of chemistry and coatings technology in Krefeld in 1990. She joined the applied technology group of Tego Chemie Service in Essen, Germany. She was responsible for various additive groups. For many years she was responsible for the Technical Service for printing inks and RC formulations. Today she is Director of the Technical Service Coatings and Inks of EVONIK Tego Chemie GmbH. She is author or coauthor of numerous papers and patents.

Katia Studer completed her PhD thesis in 2004 under the direction of Prof. Decker/University of Mulhouse. The same year, she joined Ciba Specialty Chemicals as a Post Doc and got a permanent position in 2005 as technical specialist in the Application Centre Paints & Coatings. Since April 2007, she is responsible for Curing Technologies within the Application Centre Inks & Printing of Ciba Specialty Chemicals.

Index

Symboles

A

B

Glöckner/Jung/Struck/Studer: Radiation Curing for Coatings and Printing Inks

ISBN: 978-3-86630-907-4

I

K

L

M

N

O

P

Q

R

S

T

U

V

W

Y

Z

www.ingramcontent.com/pod-product-compliance
Lightning Source LLC
LaVergne TN
LVHW081321300926
849161LV00008B/1833